AFTER
THE ICE

● Smithsonian Books

HARPER

An Imprint of HarperCollins*Publishers*
www.harpercollins.com

AFTER THE ICE

Life, Death, and Geopolitics in the New Arctic

ALUN ANDERSON

HarperCollins books may be purchased for educational, business, or sales promotional use. For information, please write: Special Markets Department, HarperCollins Publishers, 10 East 53rd Street, New York, NY 10022.

FIRST EDITION

Designed by Suet Yee Chong

Maps © Alun Anderson and Nigel Hawtin

Library of Congress Cataloging-in-Publication Data

Anderson, Alun.
    After the ice : life, death, and geopolitics in the new Arctic / Alun Anderson. — 1st ed.
        p. cm.
    Included bibliographical references and index.

    ISBN 978-0-06-157907-3

    1. Arctic regions—Environmental conditions. 2. Arctic regions—Strategic aspects. 3. Sea ice—Arctic regions. 4. Arctic regions—Description and travel. 5. Geopolitics—Arctic regions. I. Title.
GE160.A694A53 2009
909'.0913083—dc22   2009025823

---

09 10 11 12 13 OV/RRD 10 9 8 7 6 5 4 3 2 1

CONTENTS

ANIMALS

OIL AND SHIPS

FINALE

Introduction

THE BEAR ON THE BEACH

The first polar bear that I ever saw was walking steadily along a narrow strip of beach on the south coast of Devon Island. I'd arrived in Canada's High Arctic only the day before, after flying due north for six hours from the August heat of Ottawa to the little gravel airstrip at Resolute, a place that was not even marked on my atlas back home in London. A chance invitation had brought me here: a journalist colleague of mine had been invited to join a small cruise ship to see the High North, but had to cancel at the last minute. I volunteered instead, knowing nothing of what lay before me.

At Resolute I had a quick course in how to climb into an inflatable Zodiac boat and then sped off across the bay, weaving among the ice floes, to a four-hundred-foot Russian ship that was about to leave for Ellesmere Island. Once beyond the bay, the ship ran into a huge field of ice. The captain called up the Canadian Coast Guard, and its bright red icebreaker, *Des Groseilliers*, came out to cut a path through the ice for us, and then said goodbye with a blast of its siren. We sailed on east into open blue water and sunshine with the eroded, ochre-red desert cliffs of Devon Island alongside us to the north. Only the occasional glimpse of the ice cap far inland told me that I was in the Arctic and not on a cruise up the Nile.

Jet lag kept me up far into the endless light of the Arctic night, and to cure it I went up on deck. That is where I saw my first bear, and I was thrilled to pieces. She was just a white dot to begin with, but as we drew closer I could see her well with the naked eye. For the next half hour, we cruised alongside her and I could follow the bear's steady, swinging gait and watch her every move.

She wasn't exactly how I had imagined a bear would be. Her hind-quarters were higher and more massive and her neck was longer and more powerful. Most striking of all was her purposefulness. She was striding along the beach, her head stretched out in front of her, going somewhere that had nothing to do with any watching human. She never gave our ship a glance. Traveling with her, I passed into the bear's world. There are so many wild creatures that just flee at the first sight of humans, so many that you can only hope to see if you hide and keep silent. This bear looked as though she owned the beach.

I didn't know that this bear, which never even looked my way, was going to send my life in a new direction that would lead, after a few twists and turns, to this book. Perhaps that would never have happened if I hadn't begun talking to another passenger who had come up on deck, a wildlife biologist from Canada. He'd begun looking intently at the bear through a powerful tripod-mounted telescope and invited me to take a really close-up look. "She won't make it," he said casually. Make what? "Make it through the year. I've seen her several times this season walking back and forth. She's starving. She should be fat and plump now. Look at her underside and her hind legs; the fur is hanging loose. It's too late for her unless some very good luck comes along."

For a moment I felt angry: this was my very first bear. He explained that she was probably two years old and coming up to her first summer without her mother to protect her. She had not eaten enough in the spring and early summer when there had been plenty of young seals out on the ice. Perhaps she had not been quick enough to learn from her mother how to hunt. Perhaps she had been abandoned too early. Perhaps the ice had vanished too quickly. "She's walking the beach with a purpose," he said. "She's stretching out her neck, sniffing the air hoping to catch the scent of carrion. If she's lucky she'll find a dead seal washed up, or even better a whale—there are beluga and narwhal around here. If she finds it first, that is. She's young and if a bigger animal gets there before her, she'll be driven away. If there were still ice around she might be able to swim out to it and catch a seal. But we haven't seen ice since we left Resolute."

The blue sea, the beach, and the bear started to look very different. "Why isn't there ice here?" I asked. I was wondering if this was a part of the big Arctic melt. Like everyone else I'd heard stories that the Arctic ice was shrinking and seen pictures of a forlorn bear perched on a tiny ice floe in a bright blue sea. "It's not as simple as that," he replied. "This bay was full of ice last year and the year before. I'm not sure if the ice is really melting away because the world is growing warmer or this is just a bad year." If I wanted to know more about ice, he said, I should go to the ship's bridge. "They are the ones who worry about ice all voyage long. Ask them to show you some ice charts."

So I did. On the bridge there was little time to watch polar bears. The crew and the ice pilot, a local brought on board to help navigate safely through the Arctic waters, were surrounded by radar screens and weather maps. Yes, they had plenty of charts of the ice cover in the region. Far out in the Canadian High Arctic, you are not really alone. Down south in Ottawa the Canadian Ice Service captures images of the Arctic from satellites and downloads maps of the ice onto the Web for everyone to see.

One chart showed the ice concentration, going from a reassuring watery blue for "ice free" through deep greens and on to a bright red for "90 percent to total" ice cover, and a dark gray for "fast ice," that is, ice frozen up and locked to the shoreline.

Another color-coded chart showed ice age, from the mauves of fragile new ice, through the greens of first-year ice and browns of second-year ice, and on to threatening bright reds. This was the multiyear ice that had passed many times through the warmth of summer without melting away and had grown stronger, harder, and thicker. Only heavy icebreakers dare venture into the brown and red zones.

Large triangles dotted the more open waters of the ice charts: "icebergs," explained the pilot. Nearer the Greenland coast, in "iceberg alley," they were everywhere. But heavy sea ice had retreated far to the north, up Nares Strait. Here, off the coast of Devon Island where we were cruising along, the charts showed no color at all: ice free. There was just a patch of greens and browns at Resolute, back where we had called up the icebreaker to take us to open water.

One of the crew pulled out charts from earlier years. Two years before, in the same week, practically all the water between Baffin Island and Greenland had been filled by a huge tongue of first- and second-year ice;

most of the ice-free waters off Devon Island where we were sailing now had been frozen right over. Was this the Arctic melt? Staring at the charts, the answer became less certain. They showed that the ice was not so much melting away as endlessly shifting—disappearing from one place and appearing in another, thickening in one bay in one year and then vanishing from it the next. I wanted to know a lot more.

The next year, in 2007, I went along to a conference on the Arctic in Washington, D.C., and there I met Douglas Bancroft, the head of the Canadian Ice Service that had provided those ice charts.[1] Or to be more accurate, I became caught up in one of his stories. Bancroft, a tall man with a neat beard that recalls his years as a warship commander, was relating how a group of adventurers had set off for the North Pole from the northern tip of Canada just as the ice had broken away and gone off in the other direction. "They skied one way all day and drifted back the other all night," he said. The explanation for their odd heroics: "Most of them were British." As the only British person at the conference, all eyes fell on me.

Bancroft gave an enthusiastic talk explaining just how dynamic and changeable the Arctic ice was and how many unpredicted events had been taking place in the north, besides the surprise for that polar expedition. In August 2005, for example, the 3,000-year-old Ayles ice shelf had unexpectedly broken up. A gigantic area of ice nine miles long and three miles wide had broken free from the most northerly part of Ellesmere Island, and started drifting around the High Arctic.

Bancroft showed a movie created from a yearlong series of satellite images of the Arctic, the first I had ever seen. With time sped up and a view from outer space that enables you to look down on the whole Arctic from above the North Pole, the Arctic seems, he said, "almost like a dynamic, breathing, living organism, moving and shifting." The ice shivered, shimmered, pulsated, and flowed throughout the Arctic. The ice charts I had seen on board the ship had come alive.

That movie was made early in 2007, before anyone knew that a great and cataclysmic change was just about to come to the Arctic, one that would remove any doubts that the ice was in danger. In the summer of that year, an enormous area of the sea ice suddenly melted away at a speed that no one had seen before or ever imagined could happen. Compared to the previous summer, an extra 625,000 square miles turned to water. That's an area four times that of California. That giant crash grabbed my attention, too, and cemented my determination to understand the reasons why it had taken place and what it might mean for the Arctic and the world.

For decades before, the Arctic's summer ice area had been slowly shrinking, oscillating around a gentle downward trend. That trend had been fast enough to make many people worried even before the great crash of 2007. If it were to continue, many scientists feared that the Arctic might be free of summer ice by the end of the twenty-first century. But other scientists thought that we might just be witnessing a cycle of natural variability, and that the downward trend would eventually reverse and the ice would grow again. The shock of that sudden collapse in the late summer of 2007 shook up all these ideas and forced scientists to think again. All of a sudden there were new predictions that the Arctic might be free of summer ice in a decade rather than a century.

A trip to Japan gave me an opportunity to talk to another Arctic researcher, Koji Shimada. I met him at the Japan Agency for Marine-Earth Science and Technology in Yokosuka, south of Tokyo, down by the shore and tucked up against a Nissan car factory. His lab runs a famous oceanographic research vessel called the *Mirai*, but that day all the ships in harbor were slab-sided, gray, windowless monsters built to export Nissan's cars. We sat around a table in his office and he explained his own views on the ice, occasionally punctuated with the expression, "Many people do not agree with me." Shimada is well known for following his own views ("Imagination is the most important thing in science," he told me), so I wasn't surprised to hear that his childhood hero had been Naomi Uemura, the great Japanese explorer who walked alone to the North Pole.

Shimada had made a totally up-to-date, post–2007-crash version of the Arctic ice movie. He had used images from AMSR-E, a Japanese microwave sensor that can see the ice from space, even in the polar night. His film was much scarier. He projected it on the wall of his office and I sat there, feeling as though I were watching a true horror movie. As the days sped by in seconds, the whole of the Arctic's ice turned into a living organism that, in a fit of madness, was tearing itself apart.

Vast areas of ice whirled around the pole and were flung out past Iceland and down into the Atlantic. A steady torrent roared down the Nares Strait between Greenland and Ellesmere Island. Huge expanses of ice that had been locked hard to the Canadian islands were suddenly fractured and smashed to pieces, then sucked into an enormous whirlpool of moving ice hundreds of miles across. This was not a "big melt." I was not watching ice gradually turn into water, but a frozen ocean rip itself to bits as a result of forces I did not understand. I asked Shimada to play the movie again, and again. "This is not variation like we have seen in the past," Shimada said to me. "This is now catastrophe."

As I watched that movie a set of questions began to form in my mind. Why is the ice so dynamic and so unlike the "frozen North" I had expected? What is going to happen next? Will the ice all disintegrate and the summer Arctic seas soon be just clear, blue water? Or can the ice recover? I wanted to know if the change was a direct result of global warming and, if so, whether efforts to cut greenhouse gas emissions could still save the ice. After a year of reading, talking, and traveling, answers to these big questions have come to occupy the heart of this book, for I could find no other way to satisfy my curiosity and deal with the frustration (which I will explain in a moment) than to settle down and try to write it all out. The reason why the ice is vanishing—the science of Arctic change—was a great labor to understand, because the explanations kept changing as the Arctic sprang new surprises. The explanations that were in vogue when I first began asking questions had been absorbed into bigger answers a year later. Now their shape has become more stable (though nature might yet spring another enormous surprise). My efforts to tackle these big scientific questions, which determine so much else about the Arctic's future, lie at the heart of this book. But they do not form its soul.

On my journeys around the Arctic I had seen ice and icebergs and many of the Arctic's unique creatures—beluga, narwhal, and bowhead whales crossing the seas; ivory gulls sneaking around a bloodied polar bear to snatch a piece of freshly killed seal; curious walrus dragging themselves up on a beach to take a closer look at me; ringed seals following me with a wary gaze from the ice; and baby murre taking desperate leaps from their cliff ledge nests as they set out into the world, with their fathers calling them shrilly from the sea below. And I had seen a great many more bears, although I still wonder about the fate of my very first and whether she had the good fortune to survive that summer.

I traveled around Svalbard, Alaska, Norway, the Canadian islands, and both coasts of Greenland. My travels took me to Inuit communities in Canada, Greenland, and Alaska, and to stories of the troubled past and the rapid political changes that were arriving as indigenous peoples sought to run their own affairs and put a colonial past behind them. I heard stories from hunters of the first time they killed a polar bear, and listened to a child whose dream was, "to be ten so I can go out hunting on the ice with my Dad."

All too often, the city folk down south forget that the Arctic is a peopled place, and are unaware of how its inhabitants live. That can lead to some serious misunderstandings, few echoes of which reach the south. Sitting by the harbor at Tasiilaq in eastern Greenland, I was treated to a long diatribe on how European animal rights campaigners and environmental groups had impoverished tiny Greenland villages whose names I had never even heard. (I guessed trouble was coming when the first question was, "Are you from Greenpeace?" to which I could honestly answer, "No.") The campaign to ban the import of seal products, intended to stop the clubbing of baby seals off Newfoundland far away to the south, had made it impossible for Inuit to sell sealskin taken in their separate hunt of adult seals, even though the income was vital for them. This great injustice to the original people of the north and the perceived insult to their way of life from ignorant southerners were all new to me, but I did agree to go out hunting seal the next day.

The soul of this book lies with the people and creatures of the Arctic. They provide its beginning and its center and make the fate of the Arctic matter. But to talk about them, I had to look at many other things too. I needed to understand the new quarrels between nations over who owned the Arctic, where their borders should lie, and whether a boom in oil, gas, minerals, and shipping would transform the economy of the Arctic as the ice melted away. To do that I had to talk to politicians and icebreaker engineers and gain an acquaintance with oil prices, rig design, undersea pipeline–laying techniques, tanker specifications, and the horror of oil spills. All these topics have their place in this book, for I wanted to see the Arctic as a whole.

This was the cause of my frustration. No matter who I spoke to, the big picture was always lacking. I discovered that there are no experts on the Arctic and no grand sources for knowledge. There are specialists in scores of academic disciplines that each deal with a tiny part of the whole, from the behavior of whales to the patterns of ocean currents. There are politicians who worry about the borders of the Arctic, geologists who focus on gathering the evidence to define them, and whales and seals that swim right over them. Lawyers debate the 104 words of Article 234 (on ice-covered areas) of the Law of the Sea, while engineers simply see the ice as a set of complex mechanical forces that their rigs and icebreakers must withstand. Environmentalists are certain that drilling for oil in the Arctic should be banned at once while indigenous people want to see some of the world's wealth come their way. For the big picture that I was after, embracing people and ice, animals and borders, oil and ships and more, there turned out to be nobody. I had to assemble what I wanted to know by talking to well over a hundred different experts, listening to almost three hundred lectures at conferences around the world, reading many scientific papers, and seeking out books that told me what explorers had seen long ago.

There is, I think, an important reason for this excess of disconnected information. Only quite recently have people begun to see the Arctic as a region in its own right. Over the centuries, it has been a last frontier for explorers racing to the North Pole or searching for a new trade route between the Atlantic and Pacific oceans. It has been a source of quick wealth for adventurers taking its whale oils, walrus ivory, fox furs, and bear skins. It has been a Cold War border, rimmed by the defensive early warning radars of the United States and the Soviet Union, crisscrossed by the secret trails of submarines hiding under the ice, and with the ever present possibility that the air would fill with nuclear missiles in an all-out nuclear strike.

But most of the time, the Arctic has just been a long, narrow white space running across the top of maps of the world. A new view of the Arctic as a region of its own, long occupied by its own people and centered on the pole, has emerged only recently. That view owes much to Inuit political activity. Inuit came out of eastern Siberia and spread right across Alaska and Canada to Greenland, long before any such nations existed. When, in the late 1970s, Inuit set up their own Circumpolar Council to represent all their people around the Arctic, regardless of which nation they now found themselves in, they were the first to make us see the way the top of the world was interconnected.

The vanishing ice has cemented the circumpolar view. Change is coming to every part of the Arctic. As the ice retreats we can see just how close are the nations that ring the pole and how similar are the issues they and all the creatures of the Arctic face. Nothing has driven the circumpolar view forward more than the International Polar Year that lasted until the spring of 2009. Thousands of scientists—natural and social—tackled the myriad issues that are needed to form a big picture of the Arctic. My worry now is not that too little is known, but that so much is known which has not been synthesized.

The aim of this book is to provide a broad sketch of the whole, so that its different parts are recognizable and in the right places, and none are lost in an excess of detail. I think that I might be the first to attempt this overambitious goal, but I think it is important to try. The Arctic is changing so fast that no one—not the scientists that study it, the politicians who want to control it, the oilmen who want to exploit it, or the indigenous people who call it home—can keep up. The only people who appear to have gone before me, with an even bigger mission, are the brave authors of a couple of travel guides to the entire Arctic. I hope that they did not find it a foolhardy endeavor.

By writing about the Arctic as a region, though, I don't want to reinforce the notion of its being a separate, distant, remote place. Nothing could be further from reality; the Arctic is ever more entangled with the south and ever more at the mercy of decisions made elsewhere, often without the slightest consideration for the top of the world.

One day in Greenland I was out on a long trip in a little boat amid cold ice floes. We stopped for lunch on a tiny islet, where I began to run around quickly in circles to restore my circulation. After coming across a ruined grave with a human skull and bones clearly visible inside it—who knows who died there—I decided my exertions might be disrespectful. I sat down quietly to eat. Raw narwhal was served in the chilly wind. My gracious Inuit host fished around in his many layers of clothing and pulled out a small bottle, saying, "I don't know if you like this but I find it really goes well with narwhal." It was a bottle of soy sauce. Two thoughts flashed through my mind. One was how connected the whole world has become, now that soy sauce is served on a lonely Arctic islet that is home only to an unnamed grave. The other was that while eating raw whale might seem exotic, the moment you add soy sauce, you realize it is just the same old sashimi that you can eat in any Japanese restaurant.

This book does not seek to make the Arctic exotic, although I found much there that was strange. The world does not need a new form of "orientalism" centered on the north. Its focus instead is mostly on the Arctic seas, rarely traveling far into the surrounding lands, in part because of the central importance of sea and ice to the northern people, ecosystems, and economy. I only stray a little when describing the lives and future of the reindeer people of Russia. Elsewhere in the Arctic, the sea provides sustenance for those who live there, but in Russia the reindeer takes the place of the whale.

In part I have focused on the seas to keep this book to a reasonable length. Had I more space and time I would have written more of the peculiar beauty of the Arctic lands,[2] especially those polar deserts where amid arid, red-brown soil and limitless horizons, tussocks of pale yellow Arctic poppy grow, their cup-shaped flowers seeking the low sun's endless circling, possessed with a fragility that seems so out of place in the harsh North. It is with sadness that I pass them by, for, with the coming warming of the Arctic, these rare deserts, lying close to the shore-bound ice of the most northerly Arctic islands, are under the greatest threat.[3] They will disappear soon, before the children of today become adults, and I have been privileged to see them.

One other issue remains, that of "The Arctic's Revenge." Did we really think that we could make so many changes to the far-off Arctic and strip it of its ice, without the Arctic biting back? If we ever did, we were foolish.

Chapter One

TURN THE WORLD ON ITS SIDE

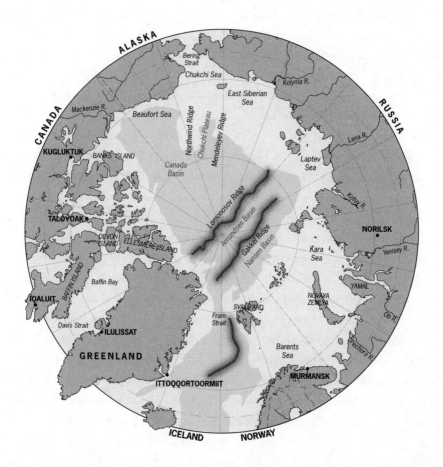

Forget the familiar view of the world with the great land masses of Asia, America, Europe, and Africa dominating the map. Instead, take hold of a globe and look straight down on it from above the North Pole. The Arctic is now laid out before you. The recognizable shapes of the great continents have mostly vanished. Your eyes are filled by a single ocean, rimmed by land. A long, smooth shoreline lies to one side and a cluster of irregular islands to the other. Only the vast ice-capped expanse of Greenland, the biggest island in the world, is immediately recognizable.

Looking down on this sea you will find much that is unfamiliar. Few of us can quickly locate the Kara or the Laptev, two of the Arctic's great seas, or the Yamal, the huge peninsula pointing out toward the North Pole where traditional reindeer herders and the Russian gas industry share the same land. The Yenisey and the Lena are two of the world's greatest rivers, each carrying more freshwater to the sea than the Mississippi or the Nile, yet they flow almost ignored into the Arctic.

The ocean at the Arctic's heart is unusual. At 5.4 million square miles, it is the smallest of the world's oceans (the Atlantic is five times its size) but 50 percent larger than the United States. Surprisingly, its closest relative is the Mediterranean, for like that much smaller sea it is hemmed in by land; not so long ago in geological time it was a lake. Now, the Bering Strait provides the Arctic Ocean with a shallow link to the Pacific, just fifty-two miles wide, while the deep Fram Strait and the Barents Sea connect it to the Atlantic. The only other route into or out of the Arctic Ocean is through the maze of narrow channels that pass among Canada's northern islands and continue to Baffin Bay and the Davis Strait.

If we could take away all the ice so that we could see right to the bottom of the Arctic seas, we would find more surprises. The shallow shelves that extend under the seas, especially from the Russian side of the Arctic, cover a little over half the ocean, more than any other in the world. Beyond the shelves, in the deeper central seas, are undersea mountain ranges, basins, and ancient plateaus, crammed together in a small space and of such baffling complexity that geologists simply don't know precisely how or when the Arctic was made. We can be sure of a few things, though. In an earlier, much warmer era, the continents of North America and Eurasia were closer together. As they gradually moved apart, the basin on the European side of the Arctic widened, leaving a split on the seafloor where molten magma welled up from deep within the earth. This flowing rock created a chain of undersea mountains, the Gakkel Ridge, which now crosses the seas on the European side of the Arctic, dividing them into the Nansen and Amundsen basins. The Gakkel Ridge has a special importance. As this split in the earth began to grow some 50 million years ago, a sliver of the Eurasian continent broke away and was left behind to form another much larger and higher chain of underwater mountains, the Lomonosov Ridge, which passes very close to the North Pole. The ridge now connects to Russia at one end and to Ellesmere Island (belonging to Canada) and Greenland (belonging to Denmark) at the other. This ancient slow-motion accident is now of great political significance. Under international treaties, whoever can provide the geological evidence that the ridge is a "natural prolongation" of their land can claim seabed rights to the ridge and a large chunk of the Arctic on either side of it too. Russians, Canadians, and Danes have all been out there busily surveying the sea bottom to gather that evidence. The Russians have even planted a flag on the seabed at the North Pole, although it is more a symbol than a threat.

The polar view of the globe quickly reveals another surprise: the truly enormous stretch of Russia's northern coast. Its Arctic lands run for over 4,000 miles—almost twice the distance between New York and San Francisco—and span eleven time zones. That distance would be larger still if Russia had not made the mistake of selling Alaska to the United States of America for two cents an acre in 1867. The true scale of Russia, the largest country in the world and not far off double the size of the United States, is hard to take in. I had my moment of realization when I visited the Russian Academy of Sciences in Moscow. A Soviet-era map of the world was on the wall, behind a symbol of Soviet power, a model of a nuclear submarine armed with rows of ballistic missiles. The map was centered so that the Soviet Union ran across the top of the world. Europe, Asia, and Africa hung down from the enormous bulk of the empire, and the Americas were relegated to a margin.

Russians live in a country that has borders with Europe at one end of their map and with Mongolia, China, Japan, and America at the other. Travel to the Inuit community living on Little Diomede Island in the Bering Strait and you can see Russia's Great Diomede Island just two and a half miles away. Russians still dream of an undersea rail tunnel linking the two continents.

Russia's vast span is important: the more you talk to Russians, the more you will realize it has shaped Russian views of the Arctic in a way that few westerners grasp. Long before the Soviet Union was created, Russian rulers were obsessed with their northern lands, partly because it was the only way they could gain access to the sea. Three hundred years ago, Peter the Great fought endless wars to get a foothold on the Baltic and Black seas to the south, but he had to build his great navy at the Arctic port of Archangel. With the Arctic so critical to the tsar, a series of "Great Northern Expeditions" were launched to map the coast and interior of Siberia. One of them, led by the Danish sea captain Vitus Bering, crossed the strait that now bears his name and reached Alaska in 1741. That's why in 1784 Russia could claim Alaska as its own, while the just-born United States had a territory that was yet to pass the Mississippi River.

The Russians moved to exploit their northern lands in a way that no other nation has attempted. Using gulag prison labor and internal exile, first under the tsars and then renewed under Stalin, towns were built across Siberia and up into the Arctic in search of minerals, timber, and other resources. Hundreds and thousands of gulag prisoners built Norilsk, the second-largest city in the Arctic after Murmansk, to exploit the region's rich nickel deposits. Further north still, the island of Novaya Zemlya also relied on prison labor. Here, more than 2,000 nuclear tests were carried out, including the detonation of a fifty-eight-megaton bomb, the largest man-made explosion in the history of the world.

No such fervor to colonize the north, regardless of expense, ever gripped the North American side of the Arctic. The first Arctic oil boom sent workers to the north slope of Alaska on a temporary basis, as though they were visiting an alien planet, rather than to settle. Even in Canada, with its 36,000 islands that give it the longest Arctic coastline of any nation, the population cling as close to the southern border as they can without calling themselves Americans. This difference has shaped the Arctic's past and will shape its future.

Look at the names on the western side of our Arctic map: Ittoqqortoormiit, Ilulissat, Iqaluit, and Kugluktuk. These Inuit names have replaced the old colonial names of Scoresby Sound, Jakobshaven, Frobisher Bay, and Coppermine. Unlike Arctic Russia, where indigenous peoples are now a small minority in their own lands, swamped by settlers from the south, a huge swath of the Arctic in Canada and in Greenland has populations that are more than 85 percent Inuit. Gradually they are taking back power, beginning with their place names, over half the Arctic. I suspect that if their strength continues to grow, the future they demand for their Arctic may surprise their southern neighbors.

Another odd feature stands out when you look at this pole-centered globe, and it turns out to be very important. All the rivers from the surrounding lands flow northward and into the Arctic Ocean. With a conventional map, which has the Arctic at its very top, you can't help thinking that the rivers flow down to the south. But they don't. All flow north, some from far-off Kazakhstan. You have to travel a long way south of the Arctic Circle to find the first rivers that flow in the other direction. This was a great annoyance to Stalin, who wanted his engineers to turn the Siberian rivers around so that they would water the arid south. In that he failed.

The Pechora, Ob, Yenisey, Kotuy, Indigirka, and Kolyma rivers of Russia, along with the Mackenzie of Canada, give 10 percent of the world's entire freshwater runoff to the Arctic Ocean, although it contains just 1 percent of the world's seawater. The effect is dramatic. Freshwater spreads out in a shallow layer on the wide shelves fringing the land, where it freezes more easily than the salty water beneath. How this freshwater mixes with the salt and how it spreads farther into the Arctic turns out to have a big influence on the cap of Arctic sea ice. And how this huge pool of freshwater eventually drains into the Atlantic may have a profound effect on the world's ocean circulation.

The rivers are critical for transport up into the Arctic, often providing the only route in for barges bearing heavy mining equipment. Once frozen, they provide some of the region's best roads: the "ice roads" used by fearless truckers. I looked for other roads on the Arctic map, but I couldn't find any. It turns out that there aren't any. Outside of Alaska's infamous Dalton Highway—the road that leads up to the northern oil fields from Fairbanks—Canada's Dempster Highway and some roads in western Russia and Scandinavia where the climate is much milder, there are very few roads in the Arctic that lead anywhere other than around settlements. This does not necessarily blunt the desire to own an automobile. Nuuk, the capital of Greenland, has more automobiles per head than anywhere outside the United States, even though it is only possible to drive around town, a favorite pastime of its residents.

Travel in the Arctic is by plane, and most routes head north–south, except for Greenland, whose only air links lead back to its colonial roots in Denmark and Iceland. Our polar map of the Arctic is thus profoundly misleading in one sense: it is a map you can't really travel around; you can only easily slide off its edges. In 1921, the great Danish-Greenlandic ethnographer Knud Rasmussen traveled over two thousand miles by dogsled along with two Inuit friends, Miteq and Arnalrulunnguaq, from the east around the edge of the Arctic and on to Siberia. He wanted to visit all the Inuit of the world and show that they were one circumpolar people. The journey took sixteen months. Ninety years later, sled is still the only way to complete this journey.

Pipelines are not on this map either, not because they don't exist, but because mapmakers don't seem to have caught up with them. I had to ferret around more than a few energy companies before I could see a complete picture (and would have to pay a small fortune to reproduce them), but they are symbolic of much of the Arctic's relation to the rest of the world. Pipelines run down from Alaska and out of Arctic Russia. Others may soon come down the Mackenzie Valley of Canada and from deep within Russia's Barents Sea. While the people of the Arctic may be cut off from one another, the rich resources of oil and gas among which they live drain away to the populated south through thousands of miles of pipe.

The map is, of course, no substitute for the real Arctic, which is a place of profound and diverse beauty. Along the shores of Devon Island in Canada's High North, great red-brown sandstone bluffs descend to the ice-filled sea. Here the Arctic is a desert. Over in Greenland, there are steep mountains and fjords and fast-flowing streams and the vast undulating plains of its great ice cap. Out on the ocean, in summer, there are just the shifting hues of the pale blue-gray sea, the passing ice floes, and a silence broken only by the breath of a passing whale, blown away by the wind while you are still wondering where it came from. All around the Arctic there is an endless rim of tundra, low-lying, covered in moss and lichen and alive with innumerable unnamed ponds and lakes and billions of hungry mosquitoes.[1] Even here, the magic is profound, such that when you return to the city, full of people, you feel a peculiar sadness.

PEOPLE

Chapter Two

IN AN INUIT LAND

I arrived at Grise Fiord on Canada's Ellesmere Island in the second week of August, which for me is high summer. Even so, the locals were surprised to see our ship. "Usually the ice doesn't go until the middle or end of August," they told us when we came ashore. "We've never seen a ship come in this early, ever."

I knew that Grise Fiord was the most northerly community in the whole of Canada, home to just 160 people. But landing that first time, I understood nothing of Inuit life, nothing of Grise Fiord's dark history.[1] I didn't yet grasp its importance in the struggle for the rights of the Inuit people, nor of the High North's connection to an Inuit leader, John Amagoalik,[2] whose work to bring self-government to the Inuit had helped create the vast new territory of Nunavut.

The houses of Grise Fiord run along the shoreline, trapped between a gravel beach, where wave-sculpted pieces of ice wash up in the surf, and the steep mountains behind. In August, veins of white snow fill the shadowed gullies in the mountainside, above a steep scree of loose rock that the winter freeze has broken from the slopes.

Approaching by sea, the eye first picks out the dull blue of a pair of giant fuel storage tanks. The wooden houses, painted white, pale blue, and shades of brown, sit on stilts to avoid contact with the frozen ground and blend into the colors of the vast landscape so well that they emerge only as the ship draws nearer.

On shore, among the houses, there is the familiar homely confusion that you'll find in any self-reliant farming community down south, where no bit of machinery, piece of old timber, box, rope, or string is thrown away because someday it might come in handy. Instead of tractors there are snowmobiles and bits of snowmobile as well as innumerable sleds that can be towed behind them, some carrying a little wooden house designed to provide shelter out on the ice.

But that is as far as the similarities to southern communities go. This is a community of hunters, not farmers. Out along the beach and sitting among the houses are the fruits of the hunt: dead animals and bits of animals left out in what is, after all, a giant freezer. The head of a walrus sits on its fat, wrinkled neck. Its dark, liquid eyes are almost closed and stare straight up at the sky. The two pure white giant tusks, fringed with delicate hairs that would have helped sense clams on the sea bottom, are now material for local carvers. The skull of a polar bear with its massive incisors lies nearby. From below someone's porch a couple of horned, furry musk ox heads look out at you with still-open eyes. Further along the beach there is the head of a narwhal, sitting in a pool of its congealed blood with its single, spiral lance pointing ten feet up into the sky. I had always wanted to see a narwhal, the inspiration for the mythical unicorn, but had never really expected my first encounter to be like this. Alongside are piles of seals, some still fresh and looking as if they might just be resting on the beach. Others have been cut up. Their gray and white dappled skins are stretched out on rectangular wooden frames and left propped up to dry on the beach while their flesh lies in bloody piles nearby.

A little later, I traveled back to Resolute, Canada's second most northerly settlement, arriving soon after a beluga whale hunt had ended. The whales, favorites of aquaria down south, were hauled up on the beach. Their white outer skin had been peeled off, leaving a red-raw body. That top layer of skin and meat is a prized delicacy. Only the white head and the beluga's famous smile was left intact. Just a couple of days earlier, I had seen hundreds of the same whales in the shallow waters of an inlet on Baffin Island. I had been sitting in a tiny boat in the shallows, watching their white shapes streak around and under my boat as they rubbed themselves along the gravel bottom to renew their pure white skin. That day was pure magic.

To meet them again, dead on the beach, was a shock, but there is more than one view of wild animals, as I gradually learned. A year or so later I was in Alaska and talking to some hunters about the tourists who come to Alaska to watch whales. "You know," one of them said, "we really don't much like people coming here to stare at our food."

Even Inuit who no longer actively hunt still crave these "country foods" and have a special attachment to them. George Edwardson, president of the Inupiat Community of the Arctic Slope, a huge stretch of the most northern part of Alaska that includes the famous whaling community of Barrow as well as many big oil developments, explained it to me this way: "The ocean is what feeds us. There is fat in the animals that live in the ocean that gives them the ability to live in the cold. We have learned as a people to borrow that fat from the animals of the sea, and that has given us the power to live in this environment."[3]

Country foods and all the tools of the hunting life are obvious signs of the link between the Inuit and the land and connect back to a time, only one or two generations ago, when success at the hunt was essential to survival. So too was sharing. A successful hunter would expect to distribute the spoils of the hunt among relatives and the wider community. Sharing remains an important part of Inuit culture, I learned as I talked to hunters. The Arctic Council's *Survey of Living Conditions in the Arctic* says the same.[4] In Alaska, "most products of hunting, fishing, and gathering do not enter the market economy," it notes. "Rather, subsistence products are directly consumed by the harvesting household, given away or exchanged. Money buys snow machines, gas, and ammunition." In places where many people have jobs and few now hunt, the selling of country foods in markets, rather than their exchange through networks of friends and relations, can still be controversial. In Iqaluit, the Canadian north's largest town, there was a long debate over whether markets that sold country foods should be encouraged. The worry was that the fruit of the hunt would become "just meat" and lose its cultural value of cementing relationships among people.

On that first visit to Grise Fiord I also began to learn that this small place and its neighbor at Resolute share a darker story, one that is critical to understanding the past and the future of the Inuit. In these communities, the long relationship between the white men who ruled Canada and the Inuit hunters of the Arctic entered its final phase before a transformation in which Inuit began to take control over their own affairs. In the history of these northern settlements are some of the worst examples of colonial paternalism as well as some of the proudest testaments to Inuit *qanuqtuurniq* (resourcefulness) and *piliriqatigiinniq* (capacity to work together for a common cause). The story of Inuit survival at Grise Fiord and Resolute is now taught in northern schools to foster these and other critical values of Inuit culture.

Grise Fiord is not an old community, as I naively assumed when I landed there, but was created in 1953 as a result of a government plan to relocate Inuit into the very northernmost part of the Canadian Arctic. Some thirty-five Inuit families were told to move 1,400 miles north by ship from their settlement at Inukjuak (formerly Port Harrison) in northern Quebec, along with sixteen people who were picked up on route from Baffin Island, to two new settlements, one at Resolute on Cornwallis Island, and one farther north, at Craig Harbour on Ellesmere Island. The Craig Harbour settlement was later moved a little west to Grise Fiord.

Government records show that the move was motivated by the need to bolster Canada's claim to the High Arctic. No Canadians lived here. In the 1950s, with the Cold War in progress and U.S. early warning stations being built throughout the Arctic, the Canadian government was growing anxious about demonstrating its sovereignty. Moving Inuit up to the High North would be a cheap way to show the flag, as they would be able to live off the land and require fewer of the facilities that would have had to be provided if white people were sent.

This is not what the Inuit families were told. The government explanation was: "This is a purely voluntary migration. . . . Under this scheme Eskimos are moved from poor hunting areas to regions where game supplies and other necessities of Arctic life are more readily available." Forty years later this "purely voluntary migration" was found by a Royal Commission to have been "one of the worst human rights violations in the history of Canada." No one consulted Inuit about what was best for them. Their fear of white people kept them obedient to the orders of government officials.

John Amagoalik was just five years old when his family was moved up to Resolute. A year or so after I visited the High North, I tracked him down at the Qikiqtani Inuit Association in Iqaluit where he now works as Director of Land and Resources. He kindly gave me some time to chat on the phone about the past and future of Nunavut. Amagoalik is popularly known as the "father of Nunavut" for the twenty-five years he spent negotiating the territory's creation. "I didn't ask for the title," he says, "but it makes me proud."

He remembers vividly the relocation, the Royal Canadian Mounted Police (RCMP) officers coming to his camp to persuade his parents that they must move to a place where the hunting would be better, the long journey north by ship up the coast of Baffin Island and the terrible news that their extended families were to be split between two locations.

"They described this new place in very glowing terms, but it turned out to be just the opposite of what we were told," recalls Amagoalik, "and they agreed to two conditions that we insisted on, the first was that we would be allowed to return home if we didn't like this new place, and secondly we would all stay together as one group. Those two promises were broken even before the year was over. When we were told our families were to be separated the women were upset and crying. I remember that the dogs started howling; that always happened when there was sadness in the family, because they were very close to us."

When they arrived at their new homes, the landscape was totally unfamiliar. Game was scarce. They had the wrong clothing, the wrong hunting equipment, and many of the promised government supplies had not arrived. They had to live in tents and hunt at a desperate speed to get in stocks for the approaching winter, when temperatures would fall to -50°C. The winter was a time of unremitting horror, when shortage of food forced them to hunt out on the ice in the twenty-four-hour darkness that in this extreme northern latitude lasts from October to March.

Among the people who went to Craig Harbour from Inukjuak was a man called Paddy Aqiatusuk who was both an expert hunter and an exceptionally skillful carver.

After a year passed, Aqiatusuk (also known by his Inuit name of Akeeaktashuk) died. Forced to hunt with his nine-year-old daughter out on the ice in the winter dark, he is said to have gradually lost the will to live. Eventually he died in a slip from an ice floe. Nothing could illustrate the powerlessness of Inuit more strongly. Aqiatusuk died at a time when his carvings, sold through intermediaries, were being exhibited in Europe and America as the works of a master. One of his works was used to illustrate a Canadian 3-cent stamp the very year that he died. Abandoned and deceived in the High Arctic, struggling to feed his family, he knew little of the value being placed on his work by "Eskimo primitive art" collectors, nor would he ever have imagined that he had achieved such celebrity that *Time* magazine would carry an announcement of his death, describing the "fluent, uncluttered simplicity" of his work just under an announcement of the young actress Audrey Hepburn's wedding.[5] Museums throughout the world hold his carvings now.

Those who remained gradually overcame the difficulties that surrounded them. The year after Aqiatusuk's death, the police moved members of the Craig Harbour community, along with their tents, some thirty-five miles west to Grise Fiord where it was easier to find seals, whales, and caribou. In the 1960s houses were built and then a school. As the years passed the community gained an astonishing knowledge of their new land.[6] Hunters ranged over an area the size of Connecticut, far larger than that used by any other Inuit community, harvesting ringed seals, walrus, narwhal, beluga, polar bear, gull eggs, Arctic char, caribou, and musk ox.

The families who had gone to Resolute suffered a different but equally dismal fate. They had ended up scavenging from an air force base dump to survive; later it was revealed that letters of complaint they wrote were never delivered, money was stolen, and sexual favors were taken by those who were supposed to care for them. Amagoalik remembers the dump. "I was there almost every day," he says, "looking for newspapers and comic books. I got an early start learning to read from that. We didn't have a school in Resolute Bay then."

Many decades later the truth about the relocation came out, and the government was forced to begin a series of enquiries that eventually recommended that the survivors receive $10 million in compensation and an apology from the government. That apology proved too much at the time;

instead the government agreed to a reconciliation statement that officials "were acting with honourable intentions in what was perceived to be in the best interests of the Inuit at that time."

Other indignities that were well meant "at that time" were suffered by the Inuit population. Because Inuit names were unfamiliar and surnames were not used, Inuit were issued numbered leather disks so that they could be identified and addressed by their numbers by white officials. The carver Paddy Aqiatusuk's "Eskimo number" was E5-715, and he signed many of his works with his number, rather than with his name. For over a century, many Inuit children were taken away from their families to live in residential schools in an attempt to assimilate them quickly, just as was tried elsewhere in North America and in Australia, and some children suffered abuse there. Amagoalik was sent off to school in Churchill in 1964, but it was the tail end of the residential school era when conditions had already improved. It was a period of change and many young Inuit became involved in political action; Amagoalik remembers listening to Martin Luther King's most famous speech on a short wave radio in Resolute. In 1974 he joined the Inuit Tapirisat of Canada (ITC), an organization representing Inuit throughout Canada, and he eventually served twice as its president. Over a period of twenty-five years, negotiations with the government involving the ITC and other organizations led to the settlement of a land claims agreement for Inuit in 1993 and then to the creation of the territory of Nunavut in 1999.

"The attitude of the government at the start was very, very negative," recalls Amagoalik. "They thought that we had no right to be sitting there, negotiating anything. I remember the first time we sat down with them, all they wanted to do was to talk about hunting and fishing, and soapstone quarrying. We wanted to talk about regaining control of our lives and our land and resources. We wanted institutions of public government; we wanted to create our own territory. We wanted financial compensation, and we wanted the protection of our language and culture and our hunting rights. It took a long time to close that gap."

In the Nunavut Land Claims Agreement (NLCA) the Inuit gave up their ill-defined native title to the land in exchange for a set of carefully laid out rights and guarantees. This included rights to vast areas of land

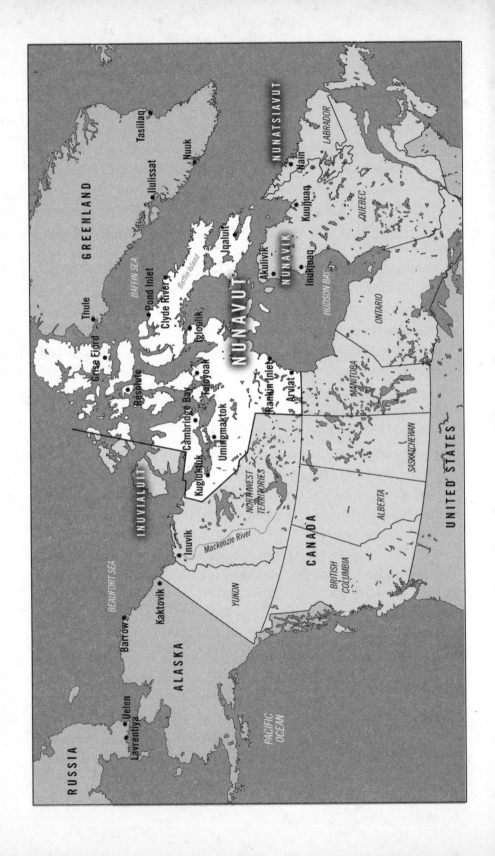

(but only a small percentage of the total); mineral rights to some promising areas for mining; three new national parks; guaranteed equal membership with the federal government in new bodies that would manage wildlife, land, and water; and over $1 billion for a fund to invest in development.

The land claim agreement is designed to compensate the Inuit and give specific rights to them as an indigenous people, unlike the government of Nunavut, which represents everyone who lives in the territory, whether Inuit (85 percent of the total) or not. This peculiarity is important; the Nunavut government and an organization called the Nunavut Tunngavik Incorporated (NTI), which looks after the Inuit's own land claim agreement, remain two powerful and separate political forces in Nunavut.

The tale of the High Arctic relocation, of Grise Fiord and Resolute, thus came close to its end in 1999 with the birth of Nunavut, but a deeper resolution was still waiting. In June 2008, soon after the Australian government apologized to its own "stolen generation," the Prime Minister of Canada, Stephen Harper, asked "the forgiveness of the aboriginal peoples of this country for failing them so profoundly. We are sorry." "It was a very emotional moment for many people," says Amagoalik.

Mary Simon, the current leader of the Inuit Tapiriit Kanatami (ITK, formerly the ITC), the organization of which Amagoalik was twice president, was in Parliament to hear the apology. I had seen photographs of her on the day, wearing a sealskin waistcoat, and had read several of her fiery political speeches, along with a piece she had written on the pleasures of berry picking, before I finally got to meet her on a cold winter's day in Quebec City in 2008. So I wasn't so surprised to find her a powerful mixture of forcefulness and warmheartedness. I asked her what the apology meant.

"It is very important. A lot of people have carried the legacy of those residential schools. An apology begins the healing process. A lot of people don't want to go to school because of the experience their parents had." In her own case, she explains that she was not sent away to school, but in her community (Kangiqsualujjuaq in Nunavik), children were not allowed to speak their own language on school grounds. "If we were caught speaking Inuktitut we were punished," she says.

And how important was the emergence of the territory of Nunavut? "It was a profound change. The majority of the Inuit live in that part of the eastern Arctic and they had always felt that the government was much too far away and didn't understand their culture and their values and their language."

At the time I met Mary Simon, the territory of Nunavut was just coming up to its tenth birthday and it had been almost fifty-five years since the High Arctic relocations. "Fifty years ago our culture was completely different and our way of life was completely different," said Simon. "We were generally afraid of the white man. We were very subservient people in many ways. That has changed. The big challenge now is for the young people to embrace their own culture and to embrace other cultures without losing their identity. We have a lot of social and economic issues to overcome."

Nunavut is not the only region of the circumpolar lands where Inuit are in the majority, but it is where the largest number of the Canadian Inuit live (49 percent of the 50,500 total). It is the only area that has gained the status of a "territory." (Canada is divided into ten provinces and three territories, with the provinces having greater devolved powers.) Elsewhere across the Canadian Inuit *Nunaat,* or Inuit homeland, there are the regions of Nunavik in northern Quebec (19 percent of the Canadian Inuit), Inuvialuit in the Northwest Territories (6 percent), and Nunatsiavut in coastal Labrador (4 percent). In all three of these areas, Inuit have succeeded in finalizing land claim agreements and have gained degrees of self-government. Nunavik is well on its way to gaining autonomy within the province of Quebec, and Nunatsiavut has already formed a regional ethnic government within its province of Newfoundland and Labrador.

Across the border from Inuvialuit, in Alaska, is the North Slope Borough, the largest "county" in the entire United States and home to 7,300 people, nearly 70 percent of them Inuit (Inupiat). Farther to the west again, around 1,600 Inuit (Yupik) remain in the Chukotka peninsula of Russia. To Nunavut's east, there is Greenland, just across Baffin Bay, with a population of 57,000, of which 90 percent are Inuit.

Nunavut is the biggest and boldest development in Inuit governance of the Arctic (although there are exciting developments in Greenland, to which we will return).[7] But the big social and economic challenges that Simon spoke of and the political difficulties of building a new territory in the Arctic cannot be exaggerated.

Arriving at the capital of Iqaluit, population 6,500, by plane from Ottawa, my first shock was the polar bear posters in the airport arrival hall advising that if attacked, you should shoot the bear and "try to get off a second shot." The word "try" remained in my mind. Out on the streets another surprise was the amazing number of young people. Downtown Ottawa is a geriatric ward in comparison. Many in the groups of young girls wearing beautiful brightly colored anoraks already had a baby tucked in its hood. This is an astonishingly young society, in which 60 percent of the population is under twenty-five. (In the United States, 32 percent of the population is under twenty-five, and in the United Kingdom it's just 28 percent.) Nunavut's population of 30,000 is growing so fast that it is expected to reach 44,000 by 2020.

The town still has a ramshackle frontier feel with dirt roads, boardwalks, white tufts of bog cotton growing wild on open bits of land, giant SUVs circling, bars where old-style brawls still occasionally break out, and a worrying number of drunks and young people begging.[8] There is a clutch of shiny new government buildings clad in blue steel, and the wonderful Nunavut Arctic College providing training in every skill from hairdressing to snowmobile repair. Iqaluit is a small town, but it is the capital of a land beyond the imagination. Everything about Nunavut's scale will leave you astonished.[9] The territory occupies 21 percent of Canada's area, or 800,000 square miles. That makes it five times the size of California and over eight times the size of Great Britain. Yet its entire population is just around 30,000 people. To put that in perspective, imagine if the entire population of the United States were just 130,000, or if fewer than 4,000 people lived in the whole of Great Britain. These nations would then have the same population density as Nunavut. Only the neighboring island of Greenland, with an equivalent land area and slightly larger population of 57,000 people, is comparable.

Nunavut's tiny population is spread out over twenty-five scattered communities. None of them are linked by roads, because there are no roads. There are no railroads, and the only deep port is for a mine at Nanisivik, on Baffin Island, that is no longer in use. Only two communities outside the capital are home to more than 1,000 people, but every small community needs its own electricity generation, school, health care facilities, and sewage and water systems. Distances between communities are vast, and air transport is expensive, infrequent, and subject to severe weather. Transport costs push the prices of daily necessities—especially bulky liquids from gasoline to orange juice—through the roof, as they do everywhere in the Arctic.

All these circumstances provide big challenges for the government of Nunavut, which has chosen to try to work in a way that will reflect Inuit culture.[10] When Nunavut came into being, it set goals of strengthening Inuit identity, delocalizing government by spreading government departments and the jobs they bring around the territory, incorporating Inuit values, working by consensus rather than conflict, and "eventually" seeking greater economic power and self-sufficiency for the territory.

The parliament began and stuck with a decision that there would be no political parties. Representatives are elected as individuals. That makes Nunavut very different from Greenland. Denmark's liberal policies through the nineteenth century created a group of educated Greenlanders very early on. By 1910, Greenland's indigenous elite had formed their own nationalist movement and were determined to end Greenlanders' status as second-class citizens compared to the Danish colonists. In contrast, Canada's Inuit were not even able to vote until the 1950s. In 1953, Greenland ceased to be a colony and became a part of the Kingdom of Denmark, and in 1979, it was granted Home Rule, a limited form of autonomy. Greenlanders gained power far earlier than Canadian Inuit, but they also modeled their democracy more closely on those of Europe. Its local parliament has parties that follow the familiar political divisions from right to left.[11]

In Nunavut, the goals of promoting and protecting Inuit cultural values are set out in the government's vision for Nunavut (*Pinasuaqtavut*) and emphasize the use of Inuit *Qaujimajatuqangit* (IQ), the valuable knowledge that Inuit have accumulated, to manage their affairs and foster Inuit so-

cial values. Among them are the following, with the first, the value of silence, perhaps most striking to those from the pushier south.

Aajiiqatigiinniq: Decision making through discussion and
consensus. Silence is part of communication and does not
necessarily signify agreement.

Inuuqatigiitsiarniq: Respecting others, maintaining relationships,
and caring for people.

Tunnganarniq: Fostering good spirit by being open, welcoming,
and inclusive.

Pijitsirniq: Serving and providing for family and/or community.

Avatittinnik/Kamatsiarniq: Respect and care for the land, animals,
and the environment.

Piliriqatigiinniq/Ikajuqtigiinniq: Working together for a common
cause.

Qanuqtuurniq: Being innovative and resourceful.

The people of Nunavut aren't alone in trying to restate their values.
In Alaska and again in Arctic Norway, I met Patricia Cochran, head of the
Inuit Circumpolar Council. She began by explaining to me that her real
name was "Sigvonna." Her mother had named her after a girl who died
young. "I was given the responsibility to carry on her life, it was an honor
and it changed me," she said. With that reminder of her different culture,
she went on to explain: "We are spending a lot of our time remembering
who we are and where we came from, because that is our strength for the
future. We have a strength that comes from our values, that comes from
our knowledge of living on the land."

That's just what the Arctic Council's Pan-Arctic SLICA survey
found.[12] Ninety percent of Inuit thought traditional activities, from berry
gathering to hunting and fishing, important to their identity. Family
ties, social support of each other, and traditional activities have a lot to
do with why indigenous people choose to remain in Arctic communities;

well-being is closely related to locally available fish and game and a sense of local control, as well as to job opportunities.

Strength is sorely needed. Appalling difficulties with education, employment, training, alcoholism, family violence, and suicide plague Nunavut as they do the whole of the Inuit land, from Russia to Greenland. "In all of our Arctic nations, in Russia too and in Greenland, in all of our communities, we have very high suicide rates, domestic violence rates, and increasing drug problems," says Cochran. "Now drugs are a very large issue in Barrow and in communities on the North Slope. Who would have ever thought that crystal meth would make its way into Arctic villages, and yet it has."

Back in Iqaluit, local statistics lend substance to this view. With fewer than 7,000 residents, police record more than 2,000 drink-related offenses each year. Across Nunavut the rate of violent crime is eight times that of Canada as a whole, unemployment reaches 50 percent in some communities, a quarter of families are headed by a single parent, only a quarter of Inuit children graduate from high school (just half the rate of other indigenous groups in Canada; the national average is closer to 90 percent), and 70 percent of adults are barely literate. Over half of Nunavut residents live in overcrowded conditions. The suicide rate in Nunavut is shockingly high. Young men between fifteen and twenty-four years of age in Nunavut are about forty times more likely to commit suicide than the average Canadian in that same age group. Although it is natural to think that high suicide rates might be linked to long, dark winters, over 80 percent of suicides occur during the time of twenty-four-hour sunlight.[13]

These are not just Inuit problems, says Cochran. "Many of the indigenous communities' worldwide face the same kind of issues because of the poverty rates, because of poor conditions that people live in, the education, and the lack of economic opportunities. I think a lot of it also has to do with the loss of culture."

Amagoalik draws attention to the particular loss of identity caused by the alleged killing of sled dogs by police officers. "The slaughtering of our dogs teams happened over a period of twenty-five years, as many as twenty-five thousand husky dogs were shot and destroyed by the RCMP during that time," he says. "Husky dogs are a very important link to our culture because they allowed us to travel and to be able to provide for our families. When that disappeared a lot of Inuit hunters who were proud hunters and proud men ended up staying at home doing nothing. I remember it very well, and the negative impact it had on our people."

The dog slaughter is very controversial. Some claim that it was a deliberate policy to end Inuit self-sufficiency and force them to switch to snowmobiles. Vehicles require gasoline, participation in a cash economy, and dependence on colonial authority. Others say that it is simply a myth and never happened at all. A massive RCMP inquiry reported in 2006 that a small number of dogs had been shot "for reasons of both public health and public safety," but found no evidence of a "widespread conspiracy to deny the Inuit people their ability to live on the land, and so force them into living in fixed settlements and a culture of perceived dependence."[14] A Truth and Reconciliation Commission has been appointed and should report in 2010. "That should really help Canada as a whole to find the truth," says Amagoalik.

Greenland, too, suffers from many of the same social problems, although the health services reflect those of the Danish welfare state. In the small town of Tasiilaq, I talked to Hans Christian Florian, the head of the local hospital and a tough mountaineer. His face became grim when he turned to the issue of young people and suicide. "The rate is very high, comparable to other Inuit communities in Canada and Alaska, but it is very difficult to give an answer to how to solve this problem. We seem not to be able to do so much about it. I think if you give young people a future or a perspective of a future, with something meaningful, education, a job, a family, I think that's what is needed, as many cases tend to be young people from a poor social background. . . . For young people here it is uphill. Where are the jobs?"

In Nunavut, the NTI blames the federal government, not the territorial government with its limited budget, for the lack of prospects. In its view, the federal government still "keeps Inuit dependent and in a state of financial and emotional despair despite promises made when the NLCA was signed in 1993."[15] At the end of 2006, the NTI sued the Government of Canada for sixteen different breaches of contract and demanded $1 billion in damages. The lawsuit is still dragging on.

The heart of the complaint is that the Canadian government failed to help train Inuit, which means that less than half the jobs in public service go to them, and they are mostly the poorly paid ones. More skilled jobs are filled by southerners.

The lawsuit did not come out of the blue. The land claim agreement did demand a "representative level of Inuit employment," and earlier talks between the NTI and government had broken down. An attempt at conciliation by Justice Thomas Berger culminated in a report in 2006 that described Nunavut as being "in a state of crisis."[16] Berger laid the blame squarely on poor education and recommended that Inuktitut, the first language for most Inuit, should be used for early education. Berger painted a graphic picture of the difficulties.

> *Imagine the odds faced by a student attempting to do homework with twelve or thirteen other people in the house (on average, half of them children), perhaps sleeping two, three, or four to a room. Nunavut's climate dictates that these tiny homes will be shut tight against the weather for possibly eight months of the year; virtually every home has at least one resident smoker; oil heating may produce carbon monoxide and other pollutants. The fact that even one-quarter of Inuit students graduate from high school is, under the circumstances, a testament to the tenacity of those students, their parents, and their communities.*

For Mary Simon, early education in Inuktitut is "a critical missing link." She explained that a relationship between early learning in one's mother tongue and later success in education had been recorded in a wide variety of cultures and societies.[17] "Unfortunately," she says, "we can teach up to grade 3 [8–9 years old] in Nunavik and other parts of the Arctic, but beyond that we don't yet have the teachers or the curriculum."

While that struggle continues, a very different challenge has arrived in Nunavut, as it has in Greenland and across the whole of the Inuit lands. Almost every Inuit community is built by the sea, and residents don't need scientists to tell them about climate change.[18]

In 2008, Amagoalik went back to his childhood home of Resolute in the summer, which he had not done for a long time. "When I left, thirty-five years ago, during the month of July, the Northwest Passage was choked with ice. There was hardly any open water. The temperature usually hovered around zero and there was very little vegetation in the surrounding islands," Amagoalik told me. "Going back, I found the Northwest Passage was ice-free, the temperature was hovering around twelve to fifteen degrees and there was a lot of new vegetation growing."

"In Nome, where I grew up," says Cochran, "there were no trees, there was no such thing as grass. I was up in Nome this summer and I saw someone not only with a lawn but mowing a lawn! I thought I would never see such a thing. There is grass, there were trees growing in yards."

Even a newcomer to the Arctic like me can spot the impact of these changes. I went for a walk in the eastern suburb of the Greenland town of Ilulissat. People don't live there; it is the area reserved for sled dogs and you can wander, very cautiously, among the fierce chained-up Greenlandic huskies. There are vacant lots and empty kennels everywhere. In 2002, 4,700 sled dogs lived here, a little more than the town's human population. Now the number is less than 3,000. The winter ice is disappearing and the locals are trading in dogs for boats.

Everyone knows change is happening, but it is harder to know where it may lead. James Ford from McGill University is one academic who has been looking at the resilience of Arctic residents. He has worked mostly in the community of Igloolik, and change has been rapid. "In the 1960s, records show that the winter ice was freezing in late September. Now it is delayed until mid-November, and the spring breakup comes earlier too," he says. Igloolik is big by Nunavut standards, with just over 1,000 residents, but it is located on an island off the Melville Peninsula and has no roads. "Ice is essential to move between communities and to hunting areas, and as a platform to hunt from," Ford explains. At first, change created difficulties for people; it was a cause of accidents and injuries, with hunters falling through thin ice and losing their equipment. But people are adapting with speed; they find new, safer routes and change hunting practices, and share new information. The hunters' ability to observe nature is helping see them through.

Amagoalik has a similar view. "Hunters are very adaptable, and if they need to change their travel patterns and if they have to learn how to hunt new species, that is what will happen," he says.

The range of animals, birds, and fish taken in the hunt can be very large as a market in Tasiilaq in Greenland reveals. Although the town has a vast store that sells almost everything you might find in Copenhagen (plus rifles and ammunition) and includes an ice cream counter, there is another small market down by the harbor. Here, *Kalaalimernit* or "wild foods" are sold rather than *Qallunaamernit* or "white man's foods." On sale are seal ($8 a kilo), polar bear ($20 a kilo), narwhal ($30 a kilo), minke whale ($18 a kilo), and walrus ($20 a kilo). The fish include salmon, cod, and halibut, and then there are many kinds of birds: murre (black guil-lemot), eider duck, kittiwake, ptarmigan, red-throated loon (red-throated diver), common loon (great northern diver), long-tailed duck, razorbill, dovekie (little auk), and seagulls, both adult and young. Seabirds like murre, razorbill, kittiwake, and dovekie are hunted in truly staggering numbers in Greenland with more than 200,000 thick-billed murre killed each year in southwest Greenland alone. At the market, razorbill are $10 each, murre $6, and dovekie, a really tiny bird, just $2 each.

Changing hunting and fishing practices is certainly possible, but hunters may need help. As the ice deteriorates, they have to travel far-ther but can't easily afford the gasoline, Ford explains. They may have to switch to fishing rather than hunting, but investing in a boat takes a lot of money. Loans may be needed, and so might insurance as hunting gear becomes more expensive. Search and rescue may need to be upgraded too. On top of these demands comes the damage to houses caused as perma-frost thaws and bigger waves in ice-free seas erode the land. The Alaskan village of Kivalina provides the most famous example. Coastal erosion means that the entire population of close to 400 people will have to be moved inland at an estimated cost of $100 million. In Alaska alone there are another twenty-six communities needing urgent action.[19] Adaptation will certainly increase the load for Arctic governments.

Nunavut's politicians are already struggling with budgets. Climate change adds to the problems of housing, unemployment, and education. Despite the great hope that greeted the birth of the new territory in 1999, it remains painfully dependent on the wealth of the south. In 2007–2008, federal payments of over C$1 billion provided around 90 percent of the Nunavut budget. Financial dependence is a bitter pill for Inuit. In Greenland the situation is similar: the population of 57,000 relies on an annual transfer from Denmark of over $500 million. Although large subsidies to less-developed peripheral regions are common in nations throughout the world, it is galling for a people that pride themselves on self-sufficiency to be dependent on others. Both Greenland and Nunavut believe their main chance of ending this dependence lies in driving development of mineral, oil, gas, and marine resources to the point where they can pay their own way.

Greenland has advanced further. In 1978, Greenlanders took responsibility only for education, church, social services, and taxation; fisheries were added later, then housing and infrastructure and, in 1992, health. In 2008 came the next big step when, in a referendum, Greenlanders voted to move toward greater independence from Denmark. The new deal recognizes Greenlandic as the official language and passes more power to Greenland. But its real importance is that it gives Greenland greater control over its natural resources and provides a formula for it to reduce its financial dependence on Denmark, step by step, as and when local revenues grow, until the point when Greenland can go it alone.

Mary Simon was one of the first to go on Canadian television to congratulate Greenlanders on voting for greater autonomy. Nunavut politicians want more control and revenues from future mining, oil, and gas operations on public land too. Currently the federal government is in charge, except for the lands specifically ceded to Inuit in the land claim agreement. The big question, of course, for both Greenland and Nunavut, is whether the natural wealth of their lands can be profitably exploited so far from southern markets in such harsh conditions. Mines have operated successfully in the past in the high Canadian Arctic at Nanisivik on Baffin Island and on Little Cornwallis Island, but they brought little benefit to the wider Nunavut economy. Still, confidence remains and whenever I asked about the future, I kept hearing one enthusiastic answer, "Mary River." That, I was told, was the kind of giant Arctic mine project that could provide wealth to Nunavut.

The very first time I heard those words I was up at Pond Inlet, in the far north of Baffin Island. Some of the locals pointed east and told me of hills of "solid iron" at Mary River that would one day make the town rich. A little later I heard another story: that one man had been pursuing the wealth of these hills for more than thirty years. It began to sound a little like a folk tale, but it wasn't.

Gord McCreary is a plain-speaking mining engineer—originally specializing in "drilling and blasting," he said, at Queens University in Ontario— who wrote an MBA thesis about Mary River some thirty years ago. It is a real place, a hundred miles inland from the north shore of Baffin Island, where there are several hills containing vast amounts of exceptionally high-quality iron ore. McCreary has not been "sitting there for thirty years waiting for my MBA topic to come around," he explained when I caught up with him by phone at his offices in Toronto, but has led a successful mining career. Still, he always believed the deposits in Baffin Island were so remarkable that one day they would be mined. The hills had been spotted from the air by the legendary exploration geologist Murray Watts. He was on his way to Iqaluit to drop off a passenger, but what he saw so excited him that he flew the 600 miles back, landed on a nearby lake, and trekked to the hill. Watts was there on July 31, 1962, says Mc-Creary, who found a signed note to that effect inside a jar buried in a cairn on top of the hill.

When I chatted with McCreary, in September 2008, he had acquired rights for Mary River and his company, Baffinland, had already spent $400 million doing everything necessary to show it was a great proposition. "We have drilled close to forty-five kilometers of cores to be sure that the ore is really there. We have had the ores tested, and we built a road to ship out a hundred-thousand-ton bulk sample to run through blast furnaces in Europe."

The ore is in four deposits spread out over a couple of kilometers. Mc-Creary thinks that there is enough iron to keep the mine active for twenty-five or perhaps even forty years. The plan is for an open cast mine at the site of "hill number one." That is not so difficult, explains McCreary, "it's just drilling and blasting, crushing and screening."

Then comes the hard part: transporting vast quantities of the ore to the blast furnaces of Europe.

McCreary's solution is a hundred miles of the most northerly rail line in the world, along with a couple of tunnels and several bridges. "Well, no, I don't characterize it as being super difficult," says McCreary, registering my astonishment. "It means that you have to know what you are building on. To mitigate the risk for rail construction you must know what your foundation materials are. So that's why we are drilling like crazy. We have as many as five drill rigs out this year, totally helicopter supported." At the end of the line, there'll be a new port. Then Baffinland will need massive icebreaking ships to shuttle the ore, summer and winter, through the Arctic ice (without disturbing marine mammals and hunters) and across the Atlantic. To achieve all this, McCreary must find big, deep-pocketed "strategic partners" with the right skills as well as sign up long-term contracts for the ore.

How much to complete the mine and to get it into production by his target date of 2013? "$4.1 billion," he told me. Baffinland is an enormous and ambitious project, but Gord McCreary, I learned, is a very serious gentleman. "I hope you appreciate," he said to me, "we are very methodical, we do this as engineers do this kind of stuff. You plan it, you work on the various components, and ultimately you get to the goal."

The potential income from the mine is enormous, McCreary believes. When I spoke to him in 2008, McCreary reckoned the value of the "proven and probable reserves" at $18 billion pretax and $11 billion posttax. "Subtract those two numbers, that's $7 billion going into the public purse," said McCreary. "And that's just for our 365 million tonnes of proven and probable reserves; we have another 500 million tonnes of resources behind that. This is bigger than a bread box, that's the way I describe it."

He might well be right—in the long run. Not so long after I spoke to him, the world economic meltdown began, iron demand slumped, and I saw his share price crash from over three dollars to fifteen cents. When I spoke to him again in 2009, he was fighting on. "We were on the point of getting our strategic partners when the world blew up. We were in discussions with some of the largest corporations in the world, and that is what it takes to move a project like this forward," he explained. "I can't put it on my American Express card."

What was it like to face an economic meltdown? "I was looking over the edge at the abyss and it was a very scary place to be," McCreary said. "But once we had made hard decisions, picked up some financing, and focused on what is important, family and things like that, over Christmas, we were feeling much more upbeat." The corporation gained that financing at a share price that made McCreary groan, but it is enough to continue work at a much reduced pace so as to be ready to roll when the recession starts to end. "I'm hopeful that we are going to get there," he says.

There are still a great many people in the High North who see the project as the way forward for the region. I found McCreary had a supporter in none other than John Amagoalik. As the Director of Land and Resources for the region, Amagoalik is negotiating the necessary Inuit Impact and Benefits Agreement (a right gained in the land claims agreement) with a view to ensuring local jobs, contracts, and the use of Inuit language in the workplace. "We've been making good progress," he says.

Nunavut's politicians worry that even these big mines, if they are built on "crown land" (federal government land) rather than Inuit-owned land, may not provide the budgets they need to develop the region. Fights between the provinces and territories and the central government over allocation of revenues are a constant theme in Canadian politics. "Right now it is not exactly fair in terms of power sharing in the resource development area," says Mary Simon. "We need to figure out a way that some of these resources can go into territorial government to improve education and health services, that the revenue sharing is done in such a way that it does not just benefit companies and businesses and southerners instead of the north."

Ottawa officials don't necessarily believe that the inhabitants of Nunavut are ready to take on extra power and responsibilities, especially the greater powers that go to a province, rather than a territory, when educational levels are low and so many Nunavut jobs are left unfilled because of lack of qualified applicants. For some in Nunavut this is a catch-22 situation: without trained people they can't get access to resources to transform the economy and train people.[20] The logic of NTI's billion-dollar lawsuit against the government for its supposed failure to honor education, training, and employment obligations becomes clear. Inuit have to find a way to break that circle.

The challenges faced by Nunavut, and Inuit all around the Arctic, are great and it would be easy to be despondent. But as Mary Simon puts it: "We always say that if other people keep making mistakes for us, as they have done, we will not succeed. If we are allowed to make our own mistakes we will learn from them." Just look at that map of the circumpolar Arctic again. The rim of one half of the entire Arctic is under some form of Inuit control now. Compare that to the situation fifty years ago, when Inuit were shuffled around the Arctic by the government. So much has been achieved. Perhaps that is the truest guide to the future.

Chapter Three

NOMADS OF THE YAMAL

F ive hours after leaving London by plane for Tokyo, you have passed Helsinki, sailed over Archangel, and sneaked around the very northern edge of the Ural Mountains, where Europe ends and Asia begins. You are now above the Yamal, a vast, low-lying tongue of land stretching 400 miles out into the Kara Sea. At its far end, like the dot at the top of an *i*, is a distinctive round island called Ostrov Belyy.

In summer, when the sun shines, the whole Yamal Peninsula sparkles with light. Everywhere braided rivers meander across its flat surface. Lakes, ponds, and streams glitter amid the green-brown of tundra. There are said to be 50,000 lakes in Yamal, but only two of them are large and even in July you can see from the air that they are still covered with ice. Passing overhead in winter, the entire peninsula is white and the sea is frozen; land and water merge and their boundary is marked only by crack lines in the shore-fast ice. Outside the plane, 30,000 feet up in the sky, the air temperature is -70°C. Down there in winter it is scarcely different: temperatures of -40°C to -50°C are made more brutal by the fierce winds blowing from the Kara Sea.

I have flown this route many times, but it is only on my most recent trip, in 2009, that I had a real appreciation of the land beneath me. The plane was routed over the southern part of Yamal and, map in hand, I could see the Yuribei River where it makes its last meandering turn on the flat tundra before flowing into the bay alongside a pointed spit of land, speckled with ponds. So that is where the pipeline is going to cross the bay, I thought. I could see no sign of life down there, but I knew that was deceptive.

Down on the Yamal are 600,000 reindeer and around 6,000 or 7,000 nomadic Nenets people who migrate with them in a yearly cycle. Their ancestors first moved to the region a thousand years ago. On their way up to join them are a railway line that will shift vast quantities of heavy equipment and the pipeline that will cross that bay and stretch 3,000 miles from the "end of the earth," as "Yamal" means in Nenets, all the way back to Germany. The enormous Bovanenkovo gas field is right in the middle of the Yamal Peninsula, and the Russian energy giant Gazprom has arrived among the reindeer.

In my earlier travels over the Yamal, I had no idea that down there were anthropologists and biologists from Russia, Germany, and America who I would one day meet. When our paths crossed at the Arctic Frontiers Conference in Tromsø in 2008, I learned of the challenge that the Nenets people of the Yamal were facing as they continued their older way of life while the largest gas development in the Arctic was taking place on the same land. I heard from them also that across the north of Russia, from the Kola Peninsula to Chukotka, other indigenous reindeer-herding groups are emerging from a turbulent past to an uncertain future. Reindeer herding is the only sustainable form of agriculture in the Russian Arctic (supplemented by fishing, berry picking, and hunting) so the fate of the herders is of special significance for the future of the Arctic.

I'd like to put my feet on the Yamal's soil too. But I don't think that is going to happen any time soon. Thanks to those gas developments, it is very hard to get a visa to fly as far as Salekhard, the nearest town to Yamal. Regular visitors include just a handful of anthropologists and biologists, and I turned to them to learn the story of the reindeer people, along with the books and papers that have been written about the Nenets. Especially I listened to two people who have been traveling to the Yamal for decades: Florian Stammler, an anthropologist from Germany, and Bruce Forbes, a biogeographer from the United States. Both now live in Finland and work at the Arctic Center of the University of Lapland in Rovaniemi, and both have a certain relaxed calm that is probably essential if you are going to live out on the tundra, sharing a *chum* (the tepeelike tent of the herders) and traveling with reindeer herders on reindeer-pulled sleds for many months. Forbes was drawn to work on the Yamal because his specialty is the impact humans have on ecosystems, while Stammler says, "I was always fascinated by nomadism."

I soon learned that for thousands of years the reindeer have meant life not just for the Nenets people of this and the neighboring regions to the east and west, but also for the Saami, Evenki, Eveny, Chukchi, and other peoples who live across the vast stretch of the Arctic from Scandinavia to the Bering Sea.[1] Including the Saami people of Fennoscandia, nearly 100,000 people from twenty indigenous groups are involved in herding 2.5 million reindeer in the circumpolar world.[2] Just as seals and whales mean life for the Inuit peoples on the other side of the Arctic, here in Siberia it is the reindeer which allow humans to survive. Only because of the reindeer's superb adaptations to extreme cold can humans live in the Russian Arctic: without the gift of this animal's meat for food, furs for warm clothing and tents, and its ability to carry people and loads across the land, there would be only a few fishers and trappers living in Siberia's river valleys and no people of the tundra. In the cold brief summers no crops can be grown, nor can any other large animal prosper.

Reindeer herding is unique to the Eurasian Arctic.[3] In North America, hunting wild caribou (as the reindeer are called there) is important to many first peoples and even vital to some, including the Gwich'in nation of the Yukon. But the people of North America never moved from hunting wild reindeer to herding them.[4] The "reindeer revolution" happened around 400 years ago and spread across Siberia with great speed. Before then, reindeer were hunted and some were tamed to carry loads and act as hunting decoys. Afterwards, people managed the herds and migrated with them, protecting them from wolves and steering them away from danger. Enormous herds of reindeer still live throughout Siberia, perhaps as many as 4 million animals in total. But many that were once herded now run wild.

Everyone agrees that the Yamal occupies center stage when considering the future of reindeer herding in the Arctic. "All the other reindeer herders in Russia defer to Yamal Nenets as the real reindeer herders who have the ultimate skills," explains Forbes. "They have retained their skills through tsarist times, through Soviet times, through the collapse of the Soviet Union, and they are still doing it, they are still sewing their own reindeer clothes." And so far, it has been a success story, with the number of reindeer doubling to about 600,000 from its low point of 300,000 in 1945.

I asked Forbes to sketch the life of a nomad reindeer herder for me and he ran through a year in the life of the Yarsalinsky, one of the groups that follow a seasonal cycle of migration along the Yamal.[5] In winter, they are found in small gather-ings south of the river Ob in the area where the tundra meets the forest. They and the reindeer take shelter here from the bitter winds. In the winter cold, the *chum*s will be covered in thick reindeer skins. There may be a meter or so of snow on the ground, through which the reindeer will dig to find lichen. Winter is a relaxed time for repairing and building sleds, sewing clothing, and fixing tents. In April, the 400-mile mi-gration to the north begins. The herders must drive their herds across the enormous expanse of the river Ob while it is still frozen, with their tents and belongings packed on sleds pulled by specially trained draft reindeer. As

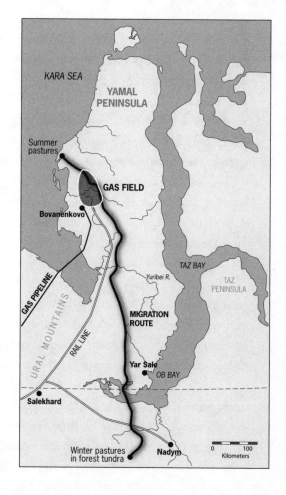

many as 550 families (almost half of all the nomadic people of the Yamal), each with their own *chum*, may be on the move from the Yarsalinsky. They are divided into twenty or so "brigades," often made up of family and relations, each of which will head for different summer pastures.

Once safely across the river, they travel up the Yamal to their small base at Yar Sale. Here they switch from winter to summer gear and leave behind the heavy reindeer skin coverings for their tents. In the past, their summer tents would have been covered in birch bark, but these days a type of canvas, bartered from oil and gas workers, has become more common. Now they travel light and fast.

The Nenets sled is tall—a little over a foot off the ground—and designed to run over both winter snow and summer tundra when knee-high shrubs may block the way. In summer, of course, the drag without the snow is much greater, so extra reindeer will be needed. The herd is in constant motion now as it races north. Every twenty-four hours or so, the herders will set up camp, sleep, wake, then break camp and quickly move on. Somewhere along the way, usually north of the Yuribei River which cuts across south-central Yamal, the pregnant does will bear their calves. The goal is to get to the coast of the Kara Sea near Kharasavei where there are onshore winds to help blow away the blackflies and mosquitoes that particularly bother the newborn. Some of the brigades will pass right through the Bovanenkovo gas field and see reindeer pastures that are now home to drill rigs.

After a few weeks on the coast where the pastures are particularly good, it will be mid-August and time to turn around and start off for Yar Sale. It is also the season for mushrooms, which reindeer love, and special care is needed to stop the herd from rushing off to search for more of these delicacies. En route those thousands of small lakes will provide fish for the herders to eat. In late September, there will be the rut, when the reindeer mate. A watch needs to be kept for wolves, which can be a threat, and for eagles, which are a danger to the young. At Yar Sale there is a modern slaughterhouse built by the Finns. Some reindeer are sold and with that, the job is done. The herders must then wait for the Ob to freeze over so that they can cross and get back to the winter pastures. Then there is time for a little rest before the cycle begins again.

In the Yamal, reindeer herding is in good shape. Many young people live out on the tundra and they often want to stay, even though the state offers inducements for them to relocate to a town. "They all know somebody living in the city, they go to Salekhard, they go to a flat and they see people drinking vodka and watching TV and committing suicide and that is not the future they want. Even the young men who go out to the military and spend two years in Chechnya come back and say they want to be a reindeer herder," says Forbes. "It is a hard life and it takes a lifetime to develop the skills to be successful, but they see it as better than stewing in an apartment in Salekhard or living on the streets in Moscow."

As an anthropologist, Stammler's relationship with the reindeer people is especially close. He has spent many seasons on the tundra, has been adopted into a herding family, and even has his own reindeer (a gift of one reindeer has led to a larger family). But he has no illusions about what it takes to be a herder.

"You need so much skill and so much intimate knowledge of the land," he says. "You must know which pasture to move to at which time of the season. You must know at what time of day, in what kinds of weather, the reindeer will prefer which kinds of lands for feeding, resting, and more. My level of expertise is minor. I can help move a household, put on the simplest type of reindeer harness, and I can slaughter a reindeer, take off the skin, fur, and intestines. These are the simplest tasks one can learn. The skill of lassoing reindeer is absolutely crucial if you want to have responsibility, but it is impossible without continuous practice."

You must also be immersed in human–reindeer relations. "Reindeer are extremely perceptive and sensitive to anything humans around them do," explains Stammler. "If you scare them off accidentally, you can create a lot of work. Reindeer can recognize individual humans, and the herders know reindeer as individuals. Relations can be very close. In one camp, a reindeer would come and visit people in the tent, although it would not come to me. It was a sacred reindeer. There is a hierarchy within the herd which is influenced by humans; in a well-managed herd you will find a master of the herd, a reindeer that has been selected for its character and may have spiritual significance for the herders."

When I first heard talk of sacred reindeers I wondered how such an apparently traditional society could possibly cope with the coming of the gas industry. I quickly learned that while the way of life might seem traditional, over the past century the Yamal herders have proved again and again their extraordinary resilience in the face of change, change that broke other reindeer-herding groups.

Before the Communists took power in 1917, the tsarist authorities were content to tax and trade and leave the indigenous people, then the clear majority in the northern lands, largely to themselves. When the Soviets arrived, no one was to be left out of the march of progress and modernization. From the 1920s, traditional religious practices were forcibly suppressed across Siberia, and many tribal shamans were imprisoned and some executed. Herding was brought under state control; villages were relocated, and work was centralized and organized around "state farms" and "production brigades" under Russian managers and political agents sent from the south. New towns and camps were built all over Siberia for migrants and for the gulag prisoners sent to exploit the region's mineral wealth. By the 1930s the indigenous people of the north had everywhere become a small minority in their homelands.

The new system was resisted in the Yamal. Herders tried to move north to get farther away from the authorities. When that failed, there were rebellions in the 1930s and 1940s, both put down forcibly. Eventually, everyone was integrated into the Soviet system and became Soviet workers, which had its advantages. Reindeer herders enjoyed the same access to education and health care as workers from anywhere else in the Soviet Union, including subsidized holidays at Black Sea beach resorts. Then, in 1991, the Soviet system suddenly collapsed, along with all the support structures that the reindeer herders had been living with for more than fifty years. "Profits" and "privatization" were words that became newly respectable. The result was disaster for many of the indigenous herders across a vast swath of Arctic Russia.

"In the post-1991 collapse of the Soviet Union," Forbes recounts, "subsidies failed in remote areas. That meant infrastructure failed, people had to hunt and concentrate on just keeping body and soul together. People were in the villages getting drunk, herds were untended and mixed with wild animals, or they were slaughtered. Hungry soldiers came along and had just vodka and ammunition. Pretty much everywhere except in the Yamal, reindeer herding contracted and in some cases completely disappeared. The people just went back to hunting."

The 400-year-old reindeer revolution was being reversed in the turmoil of the new market economy. But why were the people of the Yamal able to avoid breakdown during the "decade of chaos"? There were several special things about the Yamal Nenets, Stammler says. Throughout the Soviet period they managed to hang on to a large number of their own private reindeer, alongside the state herds they were tending. They went on living in their traditional tents and kept reindeer for transport rather than using motor vehicles which stopped running when gasoline was no longer cheaply available. And they largely kept their old migration routes.

After the collapse of Communism, they did not try to return to the way they had lived in the distant past, but retained what worked from the Soviet system. The land of the Yamal is still owned by the state, and the herders continue to organize themselves and their herding areas in three groups (the Yarsalinsky, who we've met already, the Yamalsky, and the Panaevsky) around three state farms that were set up under the Soviets and are now in the hands of local government. Only a few herders tried to cut ties with the state altogether and buy their own private land when that became possible. Instead, herders increased the size of their private herds living on state land. Under the Communist regime, 70 percent of the reindeer were state-owned and 30 percent were private. Now the percentages have reversed.

This mixture of "private" and "state" reindeer is the most bewildering aspect of the Nenets' life on the Yamal. How can you tell in a giant herd of fast-moving animals which are "private" and which are "state" reindeer? When I ask Forbes, he just laughs. Stammler says that there was a joke in post-Soviet times that "private reindeer never die." When they did, they were switched with reindeer owned by the state.

In their classic study of the Nenets, *Siberian Survival*,[6] Andrei Golovnev and Gail Osherenko recount the story of a herder of the Soviet era who had accumulated far too many private reindeer, 1,500 in all. When the

Soviet authorities came to his tent to tell him they were to be confiscated, he sent out his best-trained dog while he talked to them. The authorities departed, leaving him with just 150 reindeer. The dog then brought back 450 more animals it had moved out onto the tundra.

The Nenets have ways of getting around authority and are not slaves to tradition. "Leaders can be young, computer-savvy guys who were born in the tundra, educated in town, and who are comfortable going back and forth and going into offices. That is a set of skills that would not be available to an elder," Forbes says.

Those skills will be needed. For the past four centuries, they dealt with people who wanted their cooperation, because they wanted their reindeer meat, or to trade fish and furs. With the gas industry now moving into the Yamal, the people are not just unnecessary, they are in the way, a people who don't belong to the modern age, who "should get the picture and move to the towns," says Stammler. "There are many people in authority who still think that we can resettle these nomads," he explains. "They think they can fix the problem and don't have to mess around with agreements and consultation."

The immediate problem is that large areas of pastures are being given over to industrial development for the gas fields. Pipelines block migration routes, and junk from exploratory wells including toxic chemicals is abandoned on the tundra. All-terrain vehicles plow tracks in the tundra and litter it with smashed vodka bottles that injure reindeer, while underpaid construction workers poach reindeer for food, take fish from the lakes, and abandon dogs that run wild and harass the reindeer. Sand blows from quarries where building material has been excavated, and degradation spreads as traffic triggers melting of the permafrost.

It is not as though the state government has not noticed these issues. I heard Alexander Stotsky, deputy governor of the neighboring Nenets Autonomous District, list the problems caused by oil and gas exploration at the conference in Tromsø: "Hectic and not-organized rides of all-terrain vehicles, contamination of ground with drilling fluids, leaving industrial materials and equipment out in the tundra, intensive aviation activity harmful to the tundra fauna, and poaching."

This might sound like a classic fight between indigenous people who want to keep industry off their land and giant oil corporations that just want people out of their way. But that is all wrong. The Nenets people are not opposed to gas development; they support it. "This comes as a shock to American or Canadian people who look at the fight over whether to open Alaska's Arctic Nation Wildlife Refuge to oil development and see it as an all-or-nothing proposition," says Forbes. "Here, everybody knows that it is going ahead, they know it is for the good of the Russian state and it can help them get access to benefits like education and health care. But they also want to be able to herd reindeer, so what they need is meaningful consultation."

Sergey Serotetto, a reindeer herder from the Yamal, expressed it like this at a meeting in Nadym put together under the Arctic Council's EALÁT project: "If a reindeer herder loses his reindeer, he will lose everything he has. For him reindeer are transport, dwelling, and food. If the pastures are destroyed, the reindeer, fish, and birds will die. And they are the basis of life, not only for the Nenets but for everyone else who lives in the North. . . . We don't have to delay gas exploitation, but it must be done in a clever way in order not to damage the people who live from reindeer herding. I do not say that gas exploration must be prohibited. But what more can we lose if we lose our reindeer?"[7] The gas companies do not operate in a legal void, and over the years since the collapse of Communism, indigenous people have won some new legal rights. Much of the credit for that goes to RAIPON, the Russian Association of Indigenous Peoples of the North, which has campaigned vigorously for the recognition of traditional use of lands and rights to those lands.

Elena Andreeva of the Institute for System Analysis of the Russian Academy of Sciences in Moscow has seen all the changes of Russian life in the north. She explains that they have been partly successful, at least on paper, in gaining laws to protect the rights of indigenous people and traditional land usage (not ownership, as the land still belongs to the state) passed in 1999 and 2001. But one bill to protect reindeer herding has been under consideration for ten years, while another to strengthen indigenous peoples' rights to reject harmful development was thrown out in 2007, she says.

The reindeer herders do have one big thing going for them in that their "demands" are so moderate that no one can really object to them in principle. They want land to be reclaimed once work has been finished, old exploratory drill sites and other abandoned industrial junk to be removed, poaching to stop, industrial areas where reindeer might be harmed to be fenced off, ways across pipelines to be provided for the herds, and protected corridors left for the migrations.

No one disagrees: it is implementation that is the issue. Solving these problems is made more urgent by the first signs of climate change. Reindeer herders are adaptable and are accustomed to unpredictable shifts in the weather, but they are now beginning to encounter rivers that are still running long after they should be frozen. Sudden warm periods or rain, followed by a freeze, can create a top layer of hard ice that the herds can't break through to feed. If these events grow more common, herders will need the flexibility to move their reindeer around as changing circumstances dictate. That will be more difficult if they are hemmed in by industrial development and pipelines, and if pastureland shrinks.

The reindeer herders' future will emerge from the interaction between local government, industry, indigenous leaders, environmentalists, and, of course, the people themselves. In 2008, Stammler and Forbes took a step that few academics would be willing to take, going beyond studying the reindeer people to seeing if they could provide some practical help. "For me personally," says Stammler, "I believe you can't really do anthropology disengaged from your research partners. When you live there for such a long time you make friends, you get close to the people. You want to contribute so this place develops as best as possible, although it doesn't pay off academically. You are measured by publications and that's it."

Back at their Arctic Center at the University of Lapland, they organized a multistakeholder comanagement seminar and flew in reindeer herders from the Yamal, local officials, and representatives from Russian gas companies and pipeline companies. Not everyone showed up (some gas officials could not get permission from their bosses, and some herders were held up by a river that refused to freeze on time), but the meeting did produce a declaration of principles for development in the area in February 2009. The declaration looked for "fruitful coexistence of indigenous livelihoods and oil and gas extraction" and acknowledged "freedom of choice to lead a nomadic or sedentary way of life."[8] The hope is that the principles will help discussion among the local people, government, and oil companies, but the next steps will have to come from the local people themselves.

Forbes and Stammler remain persistently optimistic that the Nenets will find a way forward. "This could be a place where it is very possible to show that the coexistence of these two ways of using the land is achievable, given the commitment of the reindeer herders to work together with the industry and administration, and even their readiness to make sacrifices," says Stammler.

Three thousand miles away, in Chukotka, in the most easterly part of Russia, there is also some optimism for the future of reindeer herding, although this is where the "decade of chaos" hit most cruelly in a savage odyssey of abandonment, collapse, vodka, and despair.

In Soviet times, Chukotka was the number-one reindeer herding province, with 560,000 reindeer. That number fell quickly to 100,000 head. Some herds were slaughtered for quick profits in dubious business deals, while others were killed so people could survive, and many reindeer ran off and joined wild herds. The Communist ideology that gave value to indigenous people was also abandoned.

Patty Gray, now at the National University of Ireland, Maynooth, lived in Chukotka at the worst of times, working with indigenous activists and reindeer herders.[9] "The indigenous people were sent to the bottom of the heap, publicly denigrated, viscerally disenfranchised—both politically and economically—and impoverished," she says, with anger in her voice. "The particular regime that took over in 1993 was so corrupt, so belligerent, and so much of a patronage system that it excluded everybody except the favored few. It had no qualms about bankrupting the whole region."

Some in the West thought that the end of Communism would allow the herders of Chukotka to return to a romanticized, pre-Soviet idyll. "It wasn't in anyone's memory," says Gray. "The last old guys who had that knowledge died in droves in the 1990s and a lot just killed themselves because they were in such despair."

They were not the only losers in a corrupt "market economy." Many people who had migrated to Chukotka packed up and returned to the south. In 1939, before the migrants arrived, indigenous people made up 60 percent of Chukotka's population. By 1959, they were down to 25 percent, and by 1989, less than 10 percent. Now they are back up to 30 percent of the region's 58,000 inhabitants.[10]

Thankfully, there is a happier twist to this story that will delight soccer fans. Chukotka was eventually rescued by Roman Abramovich when he was elected governor in 2000. The billionaire oligarch, best known as the impassioned owner of the Chelsea Football Club in London, continued in power until 2008. Abramovich poured money into Chukotka, and the impact on its capital, Anadyr (population 12,000), has been striking, with much rebuilding taking place. In Chukotka, Abramovich is worshipped as the great hero who came and saved the people.

Virginie Vaté, an anthropologist at the Centre National de la Recherche Scientifique in Paris, has spent time recently in Chukotka, traveling with herders near Amguema and living in the traditional *yaranga* or nomadic house. "The situation has improved at least in terms of number of reindeer," she says. In Chukotka reindeer herding was more closely controlled by the state system than it was in Yamal, and the collapse was more severe. Now, once again, people are getting salaries and there are incentives for brigades to increase the number of reindeer, says Vaté. According to the Chukotka governor's office, the number of reindeer, virtually all owned by the state, has climbed back to 180,000—nowhere near the glory days of 560,000, but back on the road to recovery.

This is not quite the end of the tale. There is another message from Chukotka. Here, too, there is great natural wealth, although in the form of minerals rather than oil, but the region's terrible remoteness makes it hard to exploit them profitably. One option the Chukotka government considered just recently was for a part of the population to be encouraged to move to a more favorable climate zone. About 20,000 people would remain, 16,000 of them indigenous. Chukotka would be abandoned.

Although that idea was never implemented, it raises a wider question: when oil, gas, and mineral resources are too expensive to develop, or when, as in the case of Yamal, they are exhausted (in less than fifty years on current plans), will the future of the Arctic look more like its past? Remember that reindeer herding is the only sustainable form of agriculture in the Russian Arctic and it may prosper, especially if it not only survives the current rush for Arctic wealth, but takes what it can from it while local economies boom. That's just what a former reindeer herder who came along to the EALÁT project meeting in Nadym thought. "Academician Alferov once was asked how long oil and gas would last as an energy resource," he recounted. "His answer was about seventy years. If we hold our positions in the tundra now, the Nenets with their reindeer will reign the tundra after these seventy years have passed. And all these iron pieces will disappear in the earth."[11]

I asked Otto Habeck, who works at the Siberian Studies Center at the Max Planck Institute for Social Anthropology in Halle, Germany, and is coordinator of NOMAD, an International Polar Year (IPY) project that looks at social change among reindeer herders right across Russia's far north, for his view.

"I think that reindeer husbandry does have a future in Russia," he replied, "but it has been developing very differently in different regions over the past twenty years. In Yamal and a little bit farther to the east in the Gydan Peninsula, the herds are flourishing. In Chukotka herding came to a standstill, but is being quite successfully reestablished. But we have yet to see what the outcome is. If we turn back to the west, to the European north of Russia and the Komi Republic, the situation is more or less stable. If you look at the Kola Peninsula there is a shift from herding to ranching. There herds are rounded up a couple of times each year, but there is not the close association where herders look after their animals all day long."

Habeck acknowledges that there are risks. "The oil industry does actually pay quite a lot. So some members of a reindeer herding family may find a better income with a job, even as a security guard. If people lose the skills then there will be a problem. You cannot learn the skills from books. But oil and gas extraction is not happening in all the regions and usually has a peak after some years. I think people will see that herding can provide a source of income. So I think people will not abandon it."

What might the future look like? Habeck says that reindeer herding will always need government support. The herds were at their biggest when the Soviet Union poured money into them in the 1930s. In Europe, farming depends very much on state support, especially in the more marginal lands. Large parts of Europe would look very different if that help were not there and, over time, it has come to include support for farming's role in maintaining the landscape and rural populations. "I can imagine this will happen in Russia too," he says.

That jibes with something Bruce Forbes told me. He mentioned that the oil and gas companies might one day like to brag about having a functioning reindeer social and ecological system in the middle of a modern gas field. The reindeer herders might even become a great "brand" for the Yamal region. A few months later, on a Russian Web site I came across an advertisement for a holiday tour to Nar'yan Mar, capital of the Nenets region, which included going out to spend time with the "famous reindeer people." Maybe that is a first glimpse of what is to come. After everything the herders have been through, I hope they will be able to cope with fame and tourists.

ICE

Chapter Four

ADRIFT ON THE ICE

The strongest wooden ship that ever sailed the seas now sits in a light and spacious museum just across the harbor from downtown Oslo. Built in 1892 to a revolutionary design, the ship still takes your breath away. Deep down in its interior, rows of huge wooden buttresses, designed to brace the hull against pressure from the outside, arch upward and meet in the ship's center. Walking among them in the dim light you feel as if you have entered some ancient wooden cathedral built by trolls. Here you can best sense the vessel's enormous strength. The planking surrounding you is two feet thick and built up from layers of tough Italian oak and greenheart, a wood selected because its dense, fine grain allows it to bend without breaking. The ship's bow is backed by three huge timbers, each too big to wrap your arms around.

Immensely strong though this ship may be, it is its shape, not its bulk, which excites wonder. From whatever direction you look, from bow or stern, from high up on the galleries set in the museum walls, or from low down on the floor on which the ship's keel rests, you are struck by the smooth, endlessly rounded curve of the hull. The ship's proportions are taken from those of a coconut shell and are quite unlike those of any other boat you have ever seen, or any other boat that has ever been built, bar none.

The ship is the *Fram* and although it took place more than a hundred years ago, the story of its first great voyage remains an excellent starting point for an explanation of why the ice in the Arctic is disappearing so fast now. The *Fram* (meaning "forward") was built for the great Norwegian scientist and explorer Fridtjof Nansen. Just thirty-one years old when the

Fram was launched, Nansen was already a successful scientist and famous explorer, having completed the first crossing of the Greenland ice cap four years earlier.

At the time that Nansen conceived of the *Fram*, Westerners knew nothing of the interior of the Arctic for the simple reason that no one had ever been there. In the nineteenth century, a popular theory held that the central Arctic was open water. Ice, it was argued, formed only near land, and because warm currents flowed up toward the Arctic, the central seas should be free of ice. Some even claimed to have seen this "open Arctic sea" from the north of Greenland. Others believed that the shallow seas fringing the Arctic indicated that there was more land to be discovered, and that the North Pole might be found at the top of a mountain.

This was the state of knowledge when a newspaper article about the *Jeanette* expedition, written by the Norwegian meteorologist Henrik Mohn, caught Nansen's attention. In 1879 a U.S. expedition had left San Francisco in the *Jeanette* with thirty-three men on board, intending to find that open polar sea and continue to the North Pole. The ship had been quickly caught in the ice and was trapped for almost two years. Eventually the force of the ice was too great. The deck of the ship bowed upward as the pressure on the hull increased and the oakum packing was sent flying from between its planks. The *Jeanette* sank rapidly north of the New Siberian Islands. Twenty-five of the crew managed to reach the coast of Siberia, but twelve more, including the expedition's leader, died while trekking inland looking for help. The survivors returned home.

The newspaper article discussed an astonishing find. Wreckage from the *Jeanette*, including handwritten documents and clothes with the names of the crew on them, had apparently turned up on an ice floe off the coast of southwest Greenland, almost 3,000 miles from eastern Siberia where the ship had sunk. The wreckage must have traveled directly across the frozen Arctic Ocean and then down into the Atlantic to have made its way to Greenland. Nansen concluded that there was a flow of ice across the Arctic from the far end of Siberia.

That idea was new. There were, however, older hints that ice might travel long distances across the Arctic. In 1852, the British ship HMS *Resolute* journeyed deep into the Canadian Arctic on a fruitless hunt for the Franklin expedition, which had vanished several years earlier while searching for the Northwest Passage. During the winter of 1853, the *Resolute* became stuck in the ice off Viscount Melville Sound, far to the west of Baffin Island, and was abandoned. Two years later, the still-intact ship was found by a U.S. whaling boat in Davis Strait. It had drifted, unmanned, 1,200 miles across the Arctic in two years. The ship was repaired and returned to Queen Victoria as a gift from the United States. Later, when the ship left service, part of its timber was made into a desk and given to the U.S. government. The desk still sits in the Oval Office at the White House.

Nansen was not a man to merely contemplate a new theory. He at once set out to raise funds for an expedition to the North Pole that would, as he put it, "work with the forces of Nature and not against them." His aim was to build an immensely strong, round-hulled ship that could survive in the frozen seas by being squeezed up and out of the ice rather than being caught and crushed by it. During the summer, he would sail the ship as far north as he could off the eastern coast of Siberia where the *Jeanette* had perished, let it be frozen into the ice and then carried along across the Arctic until it was close to the North Pole. A small party would travel to the pole by dogsled and ski—at that time a revolutionary means of travel in the Arctic. His goal was both to gather new scientific information about the interior of the Arctic and to win fame by being first to the pole.

To turn this idea into reality, Nansen found a partner in the naval architect Colin Archer, a Norwegian of Scottish descent. Others were not convinced that the journey was feasible or that the flow of ice even existed. In the United States, reports that items from the *Jeanette* had been found in Greenland were dismissed as a hoax, and Nansen's plans as madness. "The history of Arctic expeditions contains enough stupidity not to also have to bear the burden of Dr. Nansen's illogical plan for self-destruction," wrote Adolphus Greely, one of America's most famous Arctic explorers.[1] "Records of unquestioned accuracy demonstrate that no system of currents exists in the North Polar Ocean," William Dall of the Smithsonian Institution in Washington, D.C., told newspapermen.[2]

Nansen was not to be discouraged. He froze the *Fram* into the ice off the New Siberian Islands in September 1893, and let it drift across the

Arctic for three years in a meandering track while he collected a huge store of new scientific observations. The journey was not easy. Weeks when the *Fram* drifted north were followed by those when it headed in the other direction. Nansen's confidence sometimes collapsed. "My theories, my castle, where I self-assuredly rose above all foolish objections, have come crashing down like a pack of cards. . . . There is no northbound current," he wrote in his diary when the ship was heading in the wrong direction.[3] But the *Fram* behaved as designed. As the ice closed in, the ship creaked and moaned and then with a bang, "jumped into the air."[4]

When the *Fram* neared the pole, Nansen and a companion left on skis with dog teams. He was right that this was the fastest way to travel across the ice, but it was still not fast enough. The two men were forced to turn back 230 miles from the pole and plod over the ice until they reached the then unmapped islands of Franz Josef Land, more than 500 miles from the mainland. They lived on polar bear and walrus through the winter and had the good fortune to run into another Arctic expedition and a ship home in the spring. Nansen hadn't reached the pole, but he had traveled farther north than anyone else. Meanwhile, the *Fram* had continued its travels and emerged with its crew safe off Spitsbergen. Despite the ridicule that his ideas had received, the Arctic ice was flowing just as Nansen had predicted.

In September 2006, a French yacht, the *Tara*, set off to repeat Nansen's voyage. The *Tara*'s wide, curving line copied the essence of the *Fram*'s unique coconut-shell design, but with a length of 118 feet and a beam of 33 feet was just a little smaller. Of course, the *Tara* was built not of wood but of aluminum, and was insulated with foam rubber rather than the reindeer hair favored by Nansen. And the *Tara* could contact base by satellite phone, while nothing was heard from the *Fram* for more than three years. The yacht went into the ice off the same New Siberian Islands and passed even closer to the pole than the *Fram*. Remarkably, the *Tara* completed the *Fram*'s three-year voyage in just sixteen months, exiting the ice off Greenland in January 2008. The great ice stream that Nansen discovered is still flowing but at double the speed of a century ago (an observation to which we will return).

Nansen's discovery of the ice stream on the eastern side of the Arctic was the first step in seeing the true, dynamic nature of the "frozen North," and would be his last big expedition. The *Fram* was to continue, taking another Norwegian polar hero, Roald Amundsen, to Antarctica, where he won the race to be first to the South Pole in 1911. Nansen went in a different direction and became a great diplomat.[5] He received the Nobel Peace Prize for his work saving the lives of thousands of refugees and prisoners in the aftermath of the First World War.

The next step, traveling even deeper inside the Arctic, had to wait for another generation of polar scientists. They took a different approach, reasoning that there was no need for a ship like the *Fram* when they could use the ice itself as a ship. Living aboard ice floe "drift camps" set up by daring pilots who would fly them deep into the Arctic in small ski-equipped planes, they would travel where the ice took them.

The first to set out on a long, uncontrollable drift was the Soviet scientist Ivan Papanin, twice honored as a "hero of the Soviet Union." In May 1937, Papanin and three colleagues landed at the North Pole, the first time that planes had touched down anywhere so far out into the Arctic. The four men, equipped with tents, food, a hand winch to sound the ocean bottom, a chess set, a dog called Jolly to warn of polar bears, a portrait of Stalin, and a wireless transmitter, set up their camp. Then they went where the ice took them, regularly broadcasting radio reports and eulogies to their great leader Stalin.[6] The whole world followed their adventures.

After six months of wandering in the central Arctic, "North Pole 1," as the ice camp was christened, began to head straight for the gap between Greenland and Spitsbergen and its speed picked up. "We continue to drift at a devilishly fast rate southward," Papanin wrote in his diary as the floe began to travel thirty miles a day. A few weeks later they grew alarmed that they were being driven at speed toward the shore and that their voyage might end "with a final smash against the coastal cliffs" of Greenland's northeast cape. Instead they traveled on farther south as the ice melted under their feet. An inflatable rubber boat became essential to move around as lakes formed on the floe. Planes could no longer land anywhere nearby. "We live as if on a powder keg," wrote Papanin, "for any moment an ice jam may occur, the ice floe crack apart and capsize, and draw us down with it to the depths." A rescue ship arrived just in time. The camp had been invaded by what Papanin called "black snakes." Sudden dark cracks in the ice would appear, spread horribly quickly, cut under tents and widen into fissures. They had been on the floe for 274 days.[7]

Many more Russian scientists came forward, willing to risk months or even years out on the ice. On the U.S. side, too, American scientists took to the ice. They set up camps on giant ice islands, ten to fifteen miles long, which had broken free from Ellesmere Island's ice shelves and traveled around the Arctic in style. The most famous of them, T3, wandered the Arctic from 1952 to 1974. Overall, in sheer numbers, the Soviets led the way. By the beginning of the 1990s, eighty-eight Soviet camps had clocked 105,000 miles of meandering Arctic travel at an average speed of almost four miles a day. With 44,000 depth soundings and vast numbers of weather and ocean measurements, they had transformed knowledge of the Arctic.[8]

Not all the work was innocent science. Both sides used ice floes to spy on the other. The United States wanted to move closer to the sites where the Soviet Union was testing huge nuclear bombs, and they especially wanted to test equipment for detecting submarines beneath the ice. The Soviets tried building a radar station out on the ice to warn of U.S. attacks. And in 1958, a Soviet nuclear bomber was spotted on an ice runway in the middle of the Arctic.[9] Whether spies or scientists, they left behind records of the tracks of their slow journeys; when added up they provide a remarkable and unexpected picture of the movement of the ice around the Arctic, one that will help us to understand why the ice is disappearing now.

The picture leaps from the page if we just draw the tracks of the *Fram* and the *Tara* onto a map of the Arctic, along with one of the Russian North Pole series (NP6) and America's famous T3 island's wanderings. The *Fram, Tara,* and NP6 drift deep within the Arctic almost parallel to the Siberian coast, travel past Greenland and on out into the Atlantic. The enormous American ice island, T3, started closer to the North

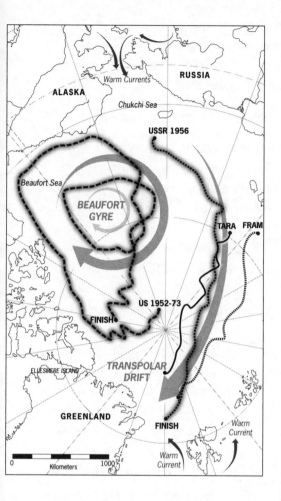

American side of the pole and simply traveled in huge circles round and round the Arctic.

The voyages chart two great streams: the Transpolar Drift Stream, running off the coast of Siberia down to Greenland, which the *Fram* followed, and the Beaufort Gyre, a slowly moving circular swirl of water some 1,250 miles across, that trapped T3. The ice island's first spin around the outer part of the gyre took almost ten years, but the second, when it had been drawn farther toward the center of this giant whirlpool, took just five years. After that, the island was flung out of the gyre and headed off toward Ellesmere Island.

Two important consequences follow. Ice in the Beaufort Gyre will have a good chance, like the ice island T3, of circling the Arctic for years. Each summer, as temperatures rise, some of the ice will melt away, but when the winter freeze returns, the surviving ice will thicken and grow. That ice will have a better chance of passing through the next melt season to grow yet again until it has become "multiyear" ice, on average a little over nine feet thick with a foot of ice sticking up above the surface of the water. This is the ice that ships fear most. As it passes through a succession of winters, the ice grows ever stronger and harder. The endlessly circling Beaufort Gyre is a giant factory that builds thick, old ice.

Ice caught in the Transpolar Drift Stream has a different fate. Like Papanin's North Pole 1 drift station, it will travel across and out of the Arctic, be devoured by "black snakes" off the coast of Greenland, and melt clean away. The Transpolar Drift Stream destroys ice.

With that picture in mind, we will leave behind the heroic age of Arctic science and move into the satellite era. Ice drift camps have not entirely disappeared from the modern age. After a long break, Russia is continuing its tradition, although in 2008, drift camp North Pole 35 had great difficulty in finding a suitable chunk of thick ice left anywhere in the Arctic. So did lots of other researchers who wanted to leave instruments on the ice, prompting many tales of scientists from different ships fighting over the best bits of ice left in the Arctic.

To chart movement of the ice now, buoys can be left out on the ice to be tracked by satellite. The cleverest of these buoys, the "Ice-Tethered Profilers," don't just sit on the ice. Beneath them, a wire rope over 2,000 feet long passes through a hole in the ice and down to the depths below. A couple of times a day, a long cylindrical package attached to the wire crawls slowly up and down its length. It is powered by a lithium-battery electric motor, "the same little electric motor that NASA picked to drive a rover around Mars," says John Toole, the scientist at Woods Hole Oceanographic Institution who designed the machine, and it is so efficient that it can "drive those instruments the distance from Boston to New York and back." Inside the package are sensors that measure the key parameters of the ocean, the temperature and the salinity at different depths. Up on top, an Iridium phone dials the laboratory at Woods Hole and sends the results to a Web site. You can just click on a map of the Arctic to see where the buoys are going and what they are finding today.

There are many other buoys out there too; gradually the Arctic is being wired as the age of autonomous exploration machines is arriving. Along with machines on top of the ice, others float freely beneath it, and there are sensors moored to the bottom of the seafloor as well as robot submarines, and underwater gliders that can travel under the ice for months on end, collecting data as they go.[10] Many of the new Arctic explorers need

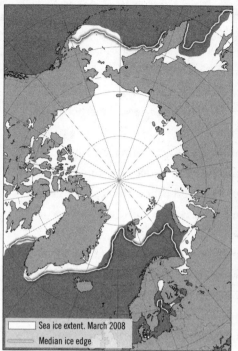

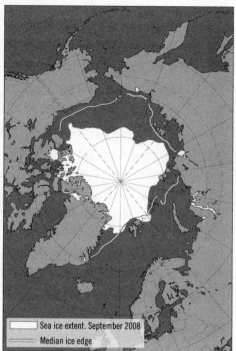

Sea ice extent. March 2008
Median ice edge

Sea ice extent. September 2008
Median ice edge

scarcely leave the comfort of the cozy labs; Nansen and Papanin would have found such ease hard to imagine.

Scientists first began to suspect that something was going wrong in the Arctic back in the 1980s. Satellite pictures showed that the area of the Arctic ice at the end of each summer was smaller than it was in the years before, although the area of ice at the end of winter scarcely changed at all. That trend was hidden amid large annual ups and downs. In 1983, the sea ice at the end of the summer covered 2.9 million square miles (7.5 million square kilometers). For comparison, the entire United States has an area of 3.5 million square miles (9.2 million square kilometers). But two years later this ice cover shrank to 2.6 million square miles (6.8 million square kilometers). The following year it rose again to just where it had been in 1983 and stayed there for another two summers. After many wobbles it was at 2.4 million square miles (6.1 million square kilometers) in 1995. Overall though, the trend was downward. The summer ice was losing about 8 percent of its area per decade and scientists needed to find an explanation.

The earth was beginning to warm and the Arctic more so: the words "greenhouse effect" and "global warming" were just entering the vocabulary. But the slightly higher temperatures that the Arctic was then seeing were not sufficient on their own to explain why the summer ice melt was gradually accelerating. Each winter, the ice returned to close to its usual area. So much ice could not be melting away in summer simply because the air was a little warmer. Something else must be going on and, in the 1990s, scientists turned to an explanation that links straight back to the discoveries made by Nansen, Papanin, and all those explorers who had followed the flow of the Arctic's ice.

Satellite maps show that the Transpolar Drift Stream and the Beaufort Gyre are not separate. The outer waters of the gyre connect to the drift stream and other routes out of the Arctic, including the Nares Strait that leads down between Greenland and Canada's Arctic islands and out through Baffin Bay and the Davis Strait. If the Beaufort Gyre tends to build ice and the Transpolar Drift Stream tends to carry it away, then it is easy to see that the balance between those two processes could be very important in determining how much ice is left in the Arctic. The two streams are driven by the winds, which are related to the pattern of the high- and low-pressure weather systems that surround the Arctic. In 1989 scientists began noticing that a pattern of winds and weather had set in that favored the export of ice out of the Arctic, rather than its retention in the Beaufort Gyre. This is the "positive phase" of the Arctic Oscillation, a pattern of atmospheric pressure differences between the Arctic and the middle latitudes. When the Arctic Oscillation enters its "negative phase," surface winds within the Arctic blow clockwise and tend to keep ice trapped in the Beaufort Gyre, where ice can grow and strengthen. In the positive phase, the clockwise winds weaken, allowing ice to slip out of the Arctic through the Transpolar Drift Stream and the Fram Strait.

The positive phase that began in 1989 was much stronger than usual and continued for seven years, right through until 1995. Nothing like it could be seen in the weather records of the previous hundred years. Normally the oscillation behaves as its name suggests and swings back and forth as the years pass. As a consequence, the ice gains in some years and loses in others. Trapped in a long positive phase, wind patterns conspired to steal ice from the Arctic and ship it off to melt in the Atlantic.

The odd behavior of the Arctic Oscillation seemed to provide an explanation for the loss of ice, but the explanation was soon found to be incomplete. The Arctic Oscillation returned to pretty much its normal behavior, although the oscillation was weaker than before, but the ice did not recover. After an exceptionally cold winter, the summer minimum in 1996 hit 3.1 million square miles (8 million square kilometers), an amount not seen since 1980. Then the downward trend continued and intensified. By 2006, the area was down to 2.2 million square miles (5.7 million square kilometers), lower than it had ever been before.

There was clearly more to the disappearance of the sea ice then a simple link to the changing wind patterns of the Arctic Oscillation. One view was that the seven long years of the positive phase of the Arctic Oscillation had stripped the Arctic of its older, thicker multiyear ice that would have been built in the Beaufort Gyre. Left with too much thin newer ice that melted more easily, and too little strong ice that could withstand the summer warmth, the Arctic ice simply could not get back on its feet. It had been hit too hard.

Then the year of the catastrophic crash arrived and shook up the whole science community. In September 2007 the summer ice shrank to just 1.7 million square miles (4.3 million square kilometers), down from 2.3 million square miles (5.9 million square kilometers) the year before and close to half of what it had been in the 1950s and 1960s. Compared to that period, an area of sea ice had been lost that was almost ten times the area of California.

Even in 2006, scientists at the U.S. National Snow and Ice Data Center had been predicting that the Arctic Ocean would have no ice in September by the year 2060.[11] Now they didn't know where the melt was leading.

Bigger explanations were needed, and the 2007 crash started a frenzy of analysis. Everyone did know that one essential part of the puzzle was incomplete. Looking down at the area of ice in the Arctic, all you are seeing is a two-dimensional picture. A three-dimensional view that shows the volume of the ice is what is needed. Thin ice will obviously melt away faster than thick ice, so if we don't know how much ice is hiding beneath the water, the future is going to be very hard to predict.

Before moving on to see what is known about ice thickness, and why it is so frustratingly difficult to obtain the data that are so important, we should register that we have found a clear suspect in the case of the disappearance of the Arctic ice. A direct link between the Arctic Oscillation and the loss of sea ice may have gone dead, but it is plain that the pattern of Arctic winds, the pattern of high- and low-pressures around the Arctic, can have a big impact on what happens to the ice and whether it stays in the Arctic or departs. The movement of the ice, its dynamic side, will remain a part of any larger explanation of the disappearance of the ice, and has another consequence which we will explore later.

We don't know nearly as much as we would like about changes in the thickness of the ice. Christian Haas at the University of Alberta in Canada has probably spent more time trying to measure the thickness of Arctic ice than anyone else on the planet. "What we are really missing as we try to understand the recent changes and improve our forecasts is ice thickness information and the means to go out there and gain that information," Haas told me at the end of 2008. The first time I talked to Haas, in late 2007, he had just moved to Canada from his home at the Alfred Wegener Institute in Germany and he had a grand plan to travel across the Arctic in a French airship. The airship would be towing his "electromagnetic [EM] bird" on a long cable so that it flew just above the surface of the ice right across the Arctic. The EM bird looks like a long cylinder with a tubular tail, but inside are electronics to generate an electromagnetic field that can measure the precise distance to the water hidden beneath the ice. A laser beam bounced off the surface of the ice measures the distance to the top of the ice, and combining the two gives a reading of the ice thickness. Haas had already flown numerous helicopter missions, but each was of limited range. The airship trip would have given Haas a thickness map of a big section across the Arctic. Sadly, it never happened. The airship was torn from its moorings in a storm while in France and crashed into a house. No one was hurt, but the project was over before it began. "It was very disappointing," says Haas.

Just a couple of years earlier, the European Space Agency launched the Cryosat satellite, equipped with a highly sensitive laser altimeter which would have been capable of measuring the height of the ice above sea level across most of the Arctic, from which the ice depth could be calculated. Seventy-six seconds after liftoff, the Russian rocket carrying Cryosat malfunctioned and the satellite plunged into the sea near the North Pole rather than orbiting above it.

"After these disasters, I began to wonder if the Arctic has some kind of curse on it—it's not supposed to be surveyed, and its secrets must never be uncovered," Haas told me. Despite the setbacks, though, he and his team succeeded in flying 12,000 miles above the ice near the North Pole, towing their EM bird behind a helicopter and later an airplane to measure the thickness of the ice.

That mission added to other methods for measuring ice thickness, many of them very ingenious and all still very incomplete. To understand them it is best to begin by going under the ice and looking upward, taking a view of the Arctic that for more than fifty years was a secret known only to the nuclear submarine fleets of Cold War navies.

Chapter Five

THE VIEW FROM BELOW

Long before scientists began to worry about what was happening to the ice, the nuclear submarines of the U.S. and Soviet navies already knew the Arctic intimately. Their attack submarines constantly prowled under the ice, testing one another's defenses, spying on naval bases, maneuvers, and bomb tests, and searching for one another. Tracking an enemy submarine without being detected required consummate skill and a detailed knowledge of the seabed's ups and downs, its hiding places, and its danger points. As the Cold War progressed, the Soviet Union developed submarines that could launch nuclear-tipped ballistic missiles, making mastery of the world under the ice of the Arctic ever more urgent. Very little that the U.S. Navy learned about the Arctic's bathymetry and the state of its ice could be passed on to scientists in those days; to reveal anything of what they had seen would be to reveal where they had been. That could only help the enemy track them down.

Submarines continue to prowl even now that the Soviet Union has become the Russian Federation. Occasionally, they go too far. In 1992, the U.S. nuclear attack submarine USS *Baton Rouge* collided with a Russian submarine just off the Russian Northern Fleet's main base at Murmansk. The Russians claimed the collision was inside their waters; the Americans denied it. Both submarines limped back to their bases, and the mission of the USS *Baton Rouge* was never revealed. A year later the USS *Grayling* hit a Russian submarine that it was trailing in the Barents Sea. This time the collision could have been more serious as the Russian vessel is believed to have been carrying nuclear-tipped ballistic missiles. Both submarines returned safely.

Big efforts have been made to create detailed but secret charts of the sea bottom. As early as 1970, a nuclear attack submarine, the USS *Queenfish*, sailed under the ice along the entire coast of the Soviet Union mapping the contours of the seabed as it went. On its journey through the frozen Laptev, East Siberian, and Chukchi seas, the submarine could rarely surface safely, although it did manage to take pictures of polar bears on the ice through its periscope. Its voyage, now almost forty years old, is one of the very few of which there is a public account.[1]

Captain Alfred McLaren records how his submarine crossed the shallow seas, navigating quietly around pingos (small hills pushed up by ice under the sea bottom) while keeping below innumerable deep-draft ice keels protruding like stalactites from the thick ice above. The submarine was almost 300 feet long, with 117 men on board, yet it sneaked under the ice with just 30 feet above its sail and 20 feet below. Among the features they found beneath them were deep ice scours running across the sea bottom, probably caused by giant icebergs. Scientists would have loved to have heard more about them, but they had to wait. McLaren records a more practical interest, noting that they could be used as "pre-launch safe havens for ballistic missile submarines."

With the minds of U.S. submariners on the serious issue of winning the Cold War, scientists who wanted to obtain submarine data from them would have to wait—more than forty years from the first submarine cruise under the ice, as it turned out. In the meantime, they turned instead to Britain's Royal Navy submarines, which made their first underice cruises in 1971. Peter Wadhams of the Scott Polar Research Institute in Cambridge, England, was on that first trip thanks to the good connections of his laboratory chief, who had been a submariner in World War II. An odd twist, though, was the source of his funding.

"My own work on board British submarines was funded largely by the U.S. Office of Naval Research," Wadhams told me. "They wanted to see submarine data, and they couldn't get hold of the classified U.S. Navy data either." Wadhams has continued to make submarine voyages to look at the ice, right up to his most recent trip across the Arctic in 2007.

Flying above the ice, says Wadhams, you'll see smooth, frozen seas surrounded by pressure ridges where colliding ice floes have pushed up walls of ice. Overall their pattern recalls an aerial view of the old English countryside with "ancient fields surrounded by tumbledown walls." From below, the view is different. Beneath the pressure ridges are enormous ice keels, many between 30 and 60 feet deep and some even reaching 150 feet. Their size, Wadhams explains, means that they make up about half of the volume of the ice in the Arctic even though they are totally invisible from above. The keels begin as little more than underwater rubble broken from the floes, but as time passes, currents polish, smooth, and harden them. As Captain McLaren recorded, navigating among them in shallow water is one of the toughest tasks for a submarine commander.

Submarines use sonar to "see" the ice, and multibeam sonar is now the underwater eye of choice. Many narrow beams of sound are sent out in a fan shape around the submarine, and their echoes are recorded. As the millions of sonar pings return, they are converted into a three-dimensional map of the undersurface of the ice. "This is how you get the thickness, shape, and size of all the features," explains Wadhams. A navy submarine cruising beneath the ice can map a 300-foot-wide swath of the ice above, and multiple overlapping tracks can quickly build three-dimensional maps of large regions of ice.

In the late 1980s, with several submarine cruises under his belt, Wadhams noticed a worrying trend in the data he was collecting. "I started to realize that when you went over the same ground again, the ice was thinner than it was before," he says. Before then, no one had worried about the ice in the Arctic or thought that it could be at risk. But his submarine cruises along a line between the Fram Strait and the North Pole were showing him that the ice had lost an average of a little over two and a half feet of its thickness between 1976 and 1987. An area around Greenland had lost seven feet.[2] Even though the area of the ice was yet to change dramatically when seen from above, Wadhams' early work warned that the ice was already being eaten away in that hidden third dimension.

Wadhams' most recent trip was a 1,600-mile, 11-day cruise under the ice from Spitsbergen to Alaska on board the nuclear-powered HMS *Tireless* in the summer of 2007. Sadly, that cruise ended in tragedy when an explosion in the submerged submarine killed two sailors. "It was very frightening, for the crew as well, as nothing like this had ever happened before. For the first half hour nobody knew what had happened, but they did all the right things and followed their training," Wadhams says. The accident happened near journey's end, and most of the data had already been gathered. It confirmed what he had been saying for more than twenty years: the ice is growing thinner and thinner. Although it is the shrinking area of the ice that has grabbed all the attention, the volume of ice in the Arctic is disappearing even faster. "The cruise by HMS *Tireless* shows the average thickness of the sea ice has fallen by 40 percent since the 1970s," Wadhams says.

By that time, a bigger picture had emerged that confirmed the early work of the Royal Navy. In 1998, the U.S. Navy finally agreed to release a treasure trove of old data gathered in scores of submarine voyages all over the Arctic, some of it forty years old. Already in the early 1990s, following the collapse of the Soviet Union, U.S. Navy submarines had begun welcoming scientists aboard submarines on special expeditionary cruises in the Arctic. Still, it took an extra push from then U.S. Vice President Al Gore to obtain older data along with maps of submarine tracks. Even so, the data was deliberately made less precise before release and did not cover areas close to Russia.

A team of scientists led by Drew Rothrock of the University of Washington in Seattle was the first to analyze the submarine data.[3] They showed that the sea ice thickness in the 1990s had fallen by more than four feet, from the average of over ten feet recorded some twenty to forty years earlier. The volume of ice had fallen by around 40 percent, just as Wadhams had estimated. And the changes seemed to be happening everywhere that the submarines had been. Ice was not simply thinning in one area while thickening in another, which could be a result of winds shuffling ice around.

"The submarine study that compared the measurements in the 1970s and 1980s with cruises in the 1990s was spectacular—they were the most significant and dramatic results we had," says Christian Haas. Frustratingly, they are still nothing close to enough. They only provide a few data points on a graph. They can't show what changes happened year by year or where in the Arctic they happened. What was really needed was accurate, Arctic-wide information on the changing ice thickness, preferably going back many decades to provide scientists with the clues they needed to solve the big question of why the ice is vanishing so rapidly.

Without easy answers, scientists had to figure out some ingenious ways of reconstructing the history of the ice. One group, from the University of Colorado and NASA, led by Jim Maslanik and Chuck Fowler, came up with a cunning way of taking information from several different satellites, each built for a quite different purpose, and then combining that data with information from yet other experiments designed to track buoys drifting through the Arctic.[4] They weren't able to provide direct measurements of ice thickness, but they did come up with the next best thing: how the age of the ice had changed in different parts of the Arctic. Ice age is a reasonable stand-in for thickness because older ice that has survived many summers and passed through many winters is thick. New ice, naturally enough, is thin.

I called up Jim Maslanik in his laboratory at the University of Colorado and asked him how and why they had reconstructed the past. "These different types of ice are pretty fascinating and I'd been interested in them for some time," Maslanik explained. "But back in the 1990s, there hadn't been the real imperative for studying them because we weren't seeing the big changes in the Arctic sea ice then. With hindsight we should have picked up on it earlier. As time went on, we began to think, 'Boy, it would really be good to have a way of measuring ice age.' I wanted to come up with something that would not just say that this ice had survived one or two melt seasons, but whether it had survived three or four melt seasons. That was impossible from standard satellite approaches."

If ice weren't constantly moving around the Arctic, finding the age of a particular patch of ice would have been pretty easy. You could just look at a series of satellite pictures of a particular area taken over the years and ask: "When was open water last seen here?" The year that spot froze over—the year the ice was "born"—would tell you how old the ice was now.

Scientists had access to plenty of pictures of the Arctic taken from satellites in NASA's Pathfinder series, which flew from the 1970s to 2007. Many pictures had been taken using infrared light—beyond the reddest part of the spectrum we can see—or microwave radiation, which is even further along the spectrum toward radio waves. At these wavelengths, ice and open water show up as different colors, so if ice stood still, some quick answers would have been within reach. But ice is constantly on the move.

"We were kicking around some ideas about measuring ice age," Maslanik continues. "Chuck [Fowler] said he knew of a way to measure ice movement, so maybe we could just try to see if we could estimate how long the ice survives by tracking its movement." They took data from the tracks of buoys that had been left out on the ice. There were enough of these buoys scattered about the ice that they could divide the Arctic into a grid and then work out, from their tracks, how each of those grid areas had changed position and shape over the years.

Combining charts of ice movement with the pictures of the Arctic ice taken from various satellites in some clever computer programs gave them the answer they were looking for. They could follow areas of ice as they moved around the Arctic and track which ice survived the succession of summer melts for one year, two years, three years and so on—and which did not.

Out of this work sprang a map which showed the changing age of the Arctic's sea ice, going back over several decades. "We were surprised at just how realistic the maps were. We'd combined all these different kinds of data and then put them together with the motion calculation. You know if we had sat down to write a proposal to ask for funds for this work we probably never would have got it," says Maslanik. "We would have been told that too many different errors would add up and it wouldn't work. Amazingly, it looked really good. So that gave us ice age."

That was a great first step, says Maslanik, but the really important thing people want to know is the thickness. He knew that first-year ice was pretty much between three and five feet thick. Multiyear ice, ice that survives one or more melt seasons, is typically between six and ten feet. But what about the ice that is five years old? Can you use an age estimate to provide a thickness estimate? Submarine and other ice-thickness data provided comparison points. It turned out that ice age gave a reasonable estimate of ice thickness, but as the ice grew older, the estimates grew less accurate.

Such, then, was the long and winding road by which a dozen or so scientists were able to create maps of the changing age of the ice in different parts of the Arctic and, along with it, a good idea of changing ice thickness.

The maps, published at the end of 2007, provide a grim picture of the rapid disappearance of the older, thicker ice that fits well with the broad trends seen from under the ice by submarines. In spring 1985, areas of older ice—where at least half the ice was five or more years old—covered 2.2 million square miles, or more than 36 percent of the Arctic ice. In spring 2007, that area of older ice had fallen by more than half to just 1 million square miles, 17 percent of the 2007 total area. The percentage of multi-year ice five or more years old fell from 31 percent to 10 percent, while that seven or more years old fell from 21 percent to 5 percent. The Arctic ice was clearly growing ever younger, thinner, and more vulnerable.

That data is for ice age, not true ice thickness, but the overall picture of vanishing older, thicker ice fits the all too few direct observations that have been made of ice thickness across the Arctic. Data from Haas' EM-bird, for example, shows that around the North Pole, ice that had a modal thickness of a little over 8 feet in 2001 had thinned to 6.5 feet in 2004 and to 3 feet in 2007.[5]

Scientists are very far from having a complete, detailed picture of the changing ice thickness that they need. If Cryosat 2—being built to replace the satellite designed to measure ice thickness that crashed near the North Pole—is launched successfully, that will be a big help. Still, with the evidence that we have, we can be sure that the ice is growing younger and thinner, and it is likely that ice volume is falling even faster than ice area is shrinking.

That takes us back to our big questions: what is making the ice thinner and what connection does it have to global warming? We have already explored one answer to the first part of that question. Changes in the pattern of winds can change the flow of ice around and across the Arctic and push ice out of the Arctic. If old ice were pushed out of the Arctic, the thickness of what remained would be smaller. What else is important?

This is where we start to run into differences of viewpoint. Some scientists stress that the ice is growing thinner because of changes up top, in the atmosphere. Average air temperatures in the Arctic have risen by 2.5°C (4.5°F) over the last twenty years, and summer sunshine can quickly warm the surface of the summer Arctic seas in which the ice sits and eat away at it, while winds shuffle it around the Arctic. Others stress that the ice is also melting from below, or its winter growth is being hit, as a result of heat coming up from deeper in the seas. Many believe that the ice is simultaneously under attack from above and below but are not sure of their relative importance and point out that the answer won't be the same everywhere in the Arctic.

At this point I have to smile. Scientists are trained to look objectively at all the evidence. But, being human, they are most excited about the particular topic they are studying. Ask an atmospheric scientist if the Arctic is thinning because of changes to the ocean or the atmosphere and you'll get a quick reply: "atmosphere." Changing weather, winds, air temperature, and sunshine are where we will find an explanation. Those who think the ocean has a big role turn out to be mostly oceanographers. They'll be looking at temperature changes in the sea, warm currents, and the layering of the ocean below the ice for an impact on the way ice grows.

So if you talk to an atmospheric scientist and then to an oceanographer (or vice versa), you can end up confused. Talk to a roomful of them, and your mind will be in a complete spin. So I'll take them one at a time, and start by plunging back under the ice, because that turns out to be a more controversial place to be. Straight away we encounter something very strange (at least to me): down there beneath the sea surface, the water is quite warm.

One of the very oddest things I learned about the Arctic is that its ice is sitting on top of seas containing more than enough heat to melt all the ice away many times over. On the Atlantic side, from 650 feet or so on down, there is a layer of water that has come in from the much warmer oceans to the south. That water doesn't just pour into the Arctic as though you were filling a bathtub, but flows in where it most easily can. The Atlantic water creeps in through the Barents Sea and the Fram Strait and travels around the coast of Siberia to the Laptev Sea, where some is pushed by the huge Lomonosov Ridge toward the Amundsen Basin and some travels onward, deeper into the Arctic. On the Pacific side, warm water from the Bering Sea slips in across the shallow Bering Strait and slides under the cold surface waters. What depth these waters end up at is mostly determined by their density (unless something is mixing them) and that depends on both their salinity and their temperature. Other things being equal, salty water is denser than freshwater and cold seawater is denser than warm seawater.

These deep waters are growing warmer. "From an Arctic perspective, it is an invasion of warm water coming from somewhere else. But really it's a warming of the world ocean. The warming of some other part of the ocean comes to the Arctic Ocean eventually. So that's going to continue just as global warming continues," explains Mike Steele, an oceanographer at the Polar Science Center at the University of Washington in Seattle with a long interest in the layering of the Arctic seas.

If global warming means warmer oceans and those warmer oceans invade the Arctic, then we surely have an explanation of why the ice is thinning. That at least is what I thought when I first heard about the warm waters beneath the ice, but unfortunately it is not so simple.

In order to find out why, we will start on the Atlantic side of the Arctic, as east and west are quite different. Down below the Arctic ice, there is an enormous reservoir of warm Atlantic water. But its heat is trapped. Sandwiched between the deeper Atlantic water and the top layer of the sea is a layer of cold water, fresher at the top and saltier below, that insulates the surface from the depths and stops its heat from escaping. It is called the cold halocline layer (the word "halocline" tells us that it has a gradient of saltiness).

When I expressed my astonishment that there is just a thin layer of icy cold water keeping the ice from melting away completely, Steele, who has been studying the halocline layer for more than twenty years, chuckled and said, "Yes, it is kind of cool."

At the very top of the Arctic Ocean is a layer of relatively fresh water, some 150 to 300 feet deep. This is where ice forms and, not surprisingly, the water is usually at a temperature close to freezing. This topmost, cold, fresh layer (the "surface-mixed layer" as scientists call it) is fed by the great Siberian rivers pouring cold freshwater onto the surface of the Arctic, by rain, and by the melting of ice in summer. As the water is fresh, it is light and floats on top of denser, saltier waters beneath. Deep down below are the warm, salty Atlantic waters. In between is the critical insulating cold halocline layer which is fed by the formation of sea ice. When seawater freezes in winter, salt is squeezed out of the growing ice crystals, leaving behind freshwater ice and cold, dense, briny water that spreads out below the lighter, surface-mixed layer. Polynyas, areas of the Arctic where the seas are regularly open in winter, help maintain this cold halocline layer. Within polynyas, newly frozen ice is constantly pushed away by winds and currents, exposing yet more open water to freeze. These "ice factories" and the brine they reject help fuel the cold halocline layer that spreads across the Arctic.

That creates an astonishing situation. Jean-Claude Gascard, leader of Europe's DAMOCLES project of Arctic observation put it like this: "If this cold halocline, which has been formed in winter, disappeared, there is nothing to stop ice from melting. All the water underneath, all the air masses above are well above the melting point and ice will melt very rapidly."

Where the cold halocline is thin, heat will more easily escape from the warm Atlantic layer. Peter Wadhams is sure that the warming of the Atlantic layer below the ice has been important in the area he looked at, where the halocline layer is perhaps not as thick as in the center of the Arctic. "The most rapid thinning of the ice seemed to have happened in the mid-1980s to mid-1990s, which was when we noticed a distinct warming of about a degree in the core of the Atlantic layer of the ice. That means more heat was being pumped in from the Gulf Stream—and the extra degree is enough to increase the heat flux from below. It was probably one of the leading causes of the thinning of the late 1980s and early 1990s," he says.

Closer to the North Pole, where Christian Haas took his recent measurements of ice thickness, ocean heat doesn't seem to be coming through. "In places where the top layer is only relatively thin, ocean heat may come through, however in most regions of the Arctic, once this water has submerged below this mixed layer, it doesn't have any effect on the ice anymore," he says.

Across the whole Atlantic side of the Arctic there is not enough data to be sure if the warm, deep waters are having a growing impact. Nor do we know what might happen in the future. "Thanks to the halocline this ice is not melting now," says Gascard. "The problem is what's going to happen with so much open ocean and with the halocline exposed to the storms." Storms may mix the surface waters, erode the halocline layer, and reinforce the destruction of the ice. A catastrophe may be waiting.

Over on the other side of the Arctic, not far from Alaska, where warm water enters the Arctic from the Pacific through the shallow Bering

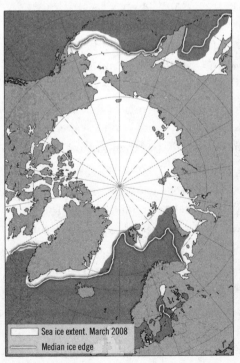
Sea ice extent. March 2008
Median ice edge

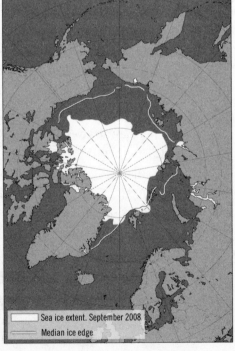
Sea ice extent. September 2008
Median ice edge

Strait, there is far more evidence that heat is already leaking from below and destroying the ice above. Here the ocean may be playing a much bigger role. This is the side of the Arctic that has seen the really massive ice losses.

Take a look again at the map of the September sea-ice minimum for 2008 and compare it to the map of the ice at the end of March when the winter ice is at its maximum. The ice hasn't shrunk back steadily from the Atlantic and Pacific sides of the Arctic toward the North Pole. Rather, it looks as though something came out of the Pacific, passed through the Bering Strait, and gobbled up all the ice within reach.

Koji Shimada, the Japanese researcher who showed me that sped-up horror movie of the Arctic's whirling ice, thinks he knows what that is, although his ideas are controversial. Under the ocean in the area where so much of the ice has gone missing, there is a large sea mount called the Northwind Ridge which is linked to the Chukchi Plateau, part of the continental shelf that fringes Alaska and Russia. In the past, Shimada explained, few ships had explored the area. The most popular route for research cruises had been between Barrow, on the northwestern tip of Alaska, and the North Pole. But more recently, trips to the area found that just below the surface there was a "very warm strong current along the Northwind Ridge, a kind of Arctic Gulf Stream."

But wasn't the cold halocline trapping heat below the surface here, too? Not really, explains Shimada.[6] On the other side of the Arctic, the Atlantic water circulates at a depth of between 1,000 and 1,300 feet, much deeper than the Pacific water; it is denser, and it is not easy for it to reach to the surface. Over on the Pacific side, the water is not strongly layered. Shimada showed me a couple of profiles of the sea around the Northwind Ridge, with bright colors indicating the temperatures at different depths. The warmth of the ocean comes right up close to the surface. "The surface water density is almost identical to the warm water. That is the key point," he says. The two waters can mix and there is no real barrier between them.

That heat has some subtle and powerful effects. "Most people who think about the reduction of the sea ice in the Arctic Ocean just think about the ice melting," he said. "It is more important to think about it as the imbalance between the ice formation that happens in winter and the ice melting that happens in summer. This is the essential issue. What happens in winter is very important." So, I needed to think beyond "faster ice melt" in the summer and include "slower ice growth" in the winter. Either way, you end up with thinner ice.

Heat from the Pacific water slows the formation of sea ice in winter, Shimada thinks, so that it ends up much thinner than that in the surrounding area. When summer comes, it is the first to disappear. That thinning of the ice triggers feedback loops. Winds can much more easily blow along a mixture of ice and open water. If the ice is really thick, winds can't do much more than whistle over its top. "If the ice moves faster, more warm water is drawn into the Arctic, which stops the winter thickening of ice," explains Shimada. "The thin ice then moves faster, drawing in more water and so on."

Sounds simple: warm water, thinner ice, winds blow the ice more easily, warm water is carried along to just where the ice is vanishing. It is not the total volume of water coming in through the Bering Strait that is important. That has not changed, explains Shimada. But the movement of the ice is growing faster, taking the warm water farther into the Arctic and up to shallower levels. This is, however, a controversial idea, with some strong supporters, mostly oceanographers, and some who think it might perhaps be a small effect in a much bigger picture. "Most atmospheric scientists don't want to understand my view," says Shimada with a sigh, thinking of that divide between those who look upward and those who look downward for explanations.

The Arctic ice itself, of course, does not give a hoot for these academic divisions, as almost anything that happens to the ice will quickly set in motion more changes, which set in motion yet others. Warm water from the Pacific might well slow the growth of winter ice and leave thinner ice that can be blown around more easily. That in turn will leave more patches of open water. And as we are about to find out, there is nothing more dangerous to an ice floe sitting out on the Arctic seas in the middle of summer than a patch of open water nearby. To wrap up the story of the vanishing ice, it is time to surface from below the seas and look at the lethal combination of sunshine and open water.

Chapter Six

THE LETHAL MIX

Sitting out on top of the ice in the middle of the Arctic Ocean are six odd-looking buoys. Each of them is about the size and shape of an automobile tire and is connected to a tangle of cables that snake out across the ice. One cable connects to a short antenna sitting alongside the buoy, another disappears down a hole in the ice, and a couple of others climb up a pole that stands nearby and looks a little like a streetlight that has landed by accident in the middle of the Arctic.

Despite their odd appearance it is easy to grow fond of these buoys. They have their own Web site where they each post their daily reports; here you can read how far they have drifted across the sea and what they have seen out there all alone in the middle of the ice. Occasionally, they come to a bad end. The six buoys sitting out there in the spring of 2009 were the survivors of a contingent of nine. The last to disappear was Buoy 2008E which "melted out" on November 14, 2008. That is, it disappeared into the sea in the vicinity of the North Pole and its data page went blank. The remaining six are still in touch and their data pages tell some good stories. These past few years, some have been seeing the emergence of dramatic feedback effects that are the last element we need to understand the great melt.

These "mass balance buoys" are put out on the ice by the U.S. Army Cold Regions Research and Engineering Laboratory (CRREL), which has its headquarters in Hanover, New Hampshire. Each of the buoys carries a pair of "pingers," one up on that pole standing close to the buoy and another hidden away beneath the ice. The pingers, or acoustic rangefinders, send out pulses of sound and record how long they take to echo back from the top and bottom surfaces of the ice. That gives a measure of the thickness of the ice and, over time, will show how quickly the top and bottom of the ice are melting away. Other instruments record the air and water temperatures.

Jacqueline Richter-Menge, an engineer at CRREL, has been looking after these buoys out on the ice since 2000 and knows them well. In the summer of 2007, the year of the ice's catastrophic disappearance from the Beaufort Sea side of the Arctic, she and her colleagues began to receive some startling messages from one of her buoys.[1]

The ice on the bottom of the floe beneath the buoy was melting away at an astonishing speed. The buoy was sitting on top of multiyear ice, usually around 3.2 meters (10.5 feet) thick, built up during the winter when air temperatures had hit -45°C (-49°F). In previous summers, when the warmth of August came around, one centimeter (0.4 inches) or so of ice would melt from the bottom of the floe each day. In 2007, the melt was averaging four centimeters a day (1.6 inches), and reached a peak of eleven centimeters (4.3 inches) a day in late August. Overall, 2.1 meters (6.9 feet) of bottom ice melted away in the summer of 2007, more than six times the annual average of 0.34 meters (1.1 feet) for the 1990s and two and a half times the melt seen during the previous summer. In contrast, the amount of melting from the surface of the ice was much the same as in previous years. Mass balance buoys in a different part of the Arctic, over near the North Pole, reported nothing unusual. "We were very surprised to see so much bottom melt in the Beaufort Sea," Richter-Menge says, "and we were suspicious about the amount of increased solar heating to the ocean too."

The summer of 2007 had been exceptionally sunny in the Beaufort Sea. Not surprisingly, Richter-Menge and her colleagues began their search for the cause of this high-speed bottom melt by calculating the impact that sunshine might have had on the ice and sea. The impact of sunshine can be counterintuitive. You might expect bright sun and clear skies to be lethal for ice, but it isn't. White surfaces have a high albedo, that is to say they are good at reflecting light, while darker ones have a low albedo and reflect little light. "If you are talking about melting ice," says Richter-Menge, "then having bright sunshine and clear days is great [that is, great for leaving the ice as it is] because the ice is a bright, high-albedo surface that reflects the sunlight back." The opposite is true for water. Seawater is a dark, low-albedo surface which soaks up sunlight and warms up fast. With open water soaking up 93 percent of the sunlight that reaches it and pure white sea ice with a layer of snow absorbing only 15 percent of the sunlight, it is open water that is really critical in warming the Arctic.

Richter-Menge and her colleagues collected data on the changing area of ice-free water and the amount of sunshine in the region. Putting those together, they could work out how much extra solar energy had been soaked up by the waters out in the Beaufort Sea. The answer was astonishing: the 2007 solar heat input in the area where their buoy was sitting was 400 to 500 percent higher than usual, mostly due to the extra areas of ocean able to soak up sunshine. The sunlight hitting the top of the ice had not had much impact. But as the warmer water washed under the ice floes, it melted them away.

The extra heat soaked up by the oceans was more than enough to account for the rapid bottom melt seen by the mass balance buoy in the Beaufort Sea ice in 2007. But what is really important is that as the ice melted, the area of dark open water grew even larger. That meant yet more solar heat was absorbed, which resulted in more ice melting, round and round in a lethal feedback loop with ice as its victim. The temperature of the sea surface just north of the Chukchi Sea rose to 4°C (39.2°F) as a huge area of ice vanished.[2] Usually the seawater here is at -1°C (30.2°F). The remaining ice in this area had no chance.

The impact of the ice–albedo feedback loop was helped along by that year's exceptionally bright, sunny weather and an unusual pattern of winds.[3] Overall the conditions in 2007 added up to what Richter-Menge describes as "the perfect storm," precipitating the catastrophic loss of ice in the summer of that year. The winds helped push ice away from the western Arctic, exposing ever more open water to soak up the bright sunshine. The pattern of winds was not a return to that positive phase of the Arctic Oscillation that had helped push ice out of the Arctic back in the 1990s. Another odd pattern has arrived instead. Jean-Claude Gascard, leader of Europe's DAMOCLES Arctic research project, explains:

"We have started to see a drastic change in the atmospheric circulation pattern. Based on the last three or so years, we are seeing a pattern with a low pressure system over Siberia and high pressure system over North America. That leaves a corridor in the center of the Arctic going from the Bering Strait on the Pacific side to the Fram Strait in the Atlantic, driving a lot of air through it." Strong winds blew straight across the Arctic. Their impact could be picked up by tracking the movement of buoys left out on the ice as part of the International Arctic Buoy program. The ice was pushed over toward the coast of Canada and Greenland,[4] leaving patches of open water behind for the bright sun to heat up. Some of that ice ended up in the Transpolar Drift Stream and vanished out through the Fram Strait.

The catastrophe of 2007 was under way, but it would never have happened if the Arctic had still had lots of its thick, old ice. Thin ice can be pushed around the Arctic by the winds much faster than can thick ice. (Remember how the *Tara* repeated the epic journey of the *Fram* in less than half the time the *Fram* took; that was because the ice is much thinner now and flowing faster with the winds.) Thin ice can also be pushed together and piled up in great crumpled piles much more easily than can thick ice. The winds of 2007 were able to sweep the ice away, leaving enough open water for the ice–albedo feedback to run away in the sunny summer, because it was so thin. As the ice was thinner than it had been before, it melted into water that much faster, speeding the feedback loop ever more.

Add together thinning ice, changing wind patterns, the albedo feedback effect, the warmth of the oceans creeping up from below, and the way they all interact with one another in a cascade of disasters, and we can try a sketch of the recent history of the Arctic ice.

The story goes like this. The slow rise in Arctic air temperatures helped gradually thin the ice over many decades. But it may not have acted alone. Warmer water flowing into the Arctic from the Atlantic during the 1980s and 1990s may have helped thin the ice from below in the eastern Arctic. Warm water from the Pacific side is slowing winter ice growth on the west. In the late 1980s to early 1990s, the unusual patterns of winds (the positive phase of the Arctic Oscillation) helped push old ice out of the

Arctic for seven years on end, leaving more of the Arctic with ice that was young, thin, and weak.

Winds, warmer air, and warmer seas added up to leave thinner ice,[5] and with less of that really strong multiyear ice, the frozen ocean became ever more vulnerable. The ice was traveling toward the disaster that arrived in the summer of 2007. That year the new pattern of winds pushed ice out of the way and bunched it up, clearing enough open water to trigger the ice-devouring albedo feedback. Months of sunny weather powered its destructive force. The "perfect storm" hit the ice of the western Arctic so hard that its area fell to a record low.[6]

It is only a sketch, but one that can neatly accommodate what happened the following year, which was nothing special at all. In the summer of 2008 nothing unusual pushed the ice farther down the road to oblivion; nor were there any exceptional events that might have helped the ice recover. It was not especially cold, nor were there wind patterns that would have trapped ice within the Arctic. Still, the area of ice left at the end of this average summer was only slightly greater than the year before, the year of the perfect storm. The ice had become too thin and vulnerable to make a comeback even when the weather was benign.

"The media didn't like that," says Mark Serreze, director of the U.S. National Snow and Ice Data Center and one of the world's best-known experts on the changing Arctic ice. "They didn't like it because it didn't break a record. The ice recovered a little bit. But you can look at that in two ways. The other way of viewing this is that even though the atmospheric pattern of the summer was not nearly as favorable to loss of ice as the previous year, we still went all the way to the second lowest ice area on record. That is frightening."

The ice may have passed a tipping point.[7] Once the ice is too thin and there is too much open water, the ice–albedo effect will kick in quickly every year and melt ever more of the vulnerable ice, leaving warmer water behind to slow the growth of winter ice. Past the tipping point, the rising temperatures due to global warming (what scientists call the "external forcing") scarcely matter anymore because the ice's own self-destructive dynamics have taken over.

With that sketch of what has been happening to the ice, it is possible to start following what is happening in scientists' minds. In a unique project scientists have been putting their thoughts on public display. In 2008, the U.S. SEARCH (Study of Environmental Arctic Change) program began asking sea ice research groups around the world to send in a monthly "outlook" giving their views on the sea ice and, if they dared, a prediction of the area of the sea ice at the end of the summer. All the ideas and predictions go up on a Web site for everyone to see, and many participants have bottles of wine wagered on the outcome.

SEARCH stresses that this is a serious program to enhance scientific cooperation and not an Internet gaming site. Just in case anyone gets confused, they publish a big, bold "DISCLAIMER," embarrassingly like those you see on dodgy stock market tip sheets, saying "This Sea Ice Outlook should not be considered as a formal forecast or prediction for Arctic sea ice extent. . . ."

In May 2008, nineteen groups sent in an "outlook." Initial forecasts ranged from 1.1 to 2.2 million square miles (2.9 to 5.6 million square kilometers) for the area of ice that remained at the end of summer. Remember that the record one-month low after the great crash of 2007 was 1.6 million square miles (4.3 million square kilometers) and the average over the previous decade had been 2.6 million square miles (6.7 million square kilometers). The high forecasts came from "conservatives" who thought sea ice would recover toward its long-term trend, the years when it was just losing a steady 10 percent of its area per decade. The low numbers came from "catastrophists" who saw a second crash hurtling toward them. As it turned out, the final 2008 summer sea-ice minimum was 1.8 million square miles (4.7 million square kilometers), a small recovery from the great crash in 2007. Only one group came close with its forecast, but the median of all forecasts was even closer, suggesting that wisdom is best found in the crowd. The best part of the game is that it enables ordinary people to follow how scientists are developing their thinking about the ice and the logic behind their forecasts.

We can't really expect perfection, as we'd need to know the weather coming to the Arctic in advance plus have a very accurate map of that year's starting conditions, which would include a map of the thickness of the ice that we can't yet make. Given the uncertainties, scientists seem to be doing well. But new links and feedbacks are still being discovered.

David Barber, a climatologist from the University of Manitoba, led a remarkable project which sent the Canadian icebreaker *Amundsen* out to the Beaufort Sea from September 2007 until August 2008, right though the winter. It spent much of its time off Cape Bathurst where a giant lead— a long, wide fracture—usually opens up in the ice. No one had ever spent the winter out in the Arctic on a research icebreaker before, so there were plenty of surprises.

"We were expecting ice to form in October, but by the end of November we still had no ice," says Barber. The first part of the explanation is familiar. In the long, hot summer of 2007 the open waters had soaked up heat and were just too warm to freeze quickly. "Climate stations on land were reporting air temperatures of -20°C, but over the oceans it was wide open and temperatures were much warmer, around -2°C, -3°C," says Barber. Then came the surprise. Huge areas of warm air above the ocean fed storms that dumped snow onto the remaining ice. "A meter of fluffy white snow built up and insulated the multiyear pack ice from the cold air so it did not grow as much as it should have done," explains Barber. That left thinner ice to face the summer melt.

So here is yet another little loop to add to the thicket of interrelated changes we have seen in the Arctic: open water warms up in the sun, warmer water can alter storm tracks, storms drop more snow on ice floes that insulates them from the cold, the ice can't grow, the ice melts away faster in summer leaving more open water, and so on. "Scientists have never had the ability to measure these kinds of processes in a detailed way before," says Barber.

When you start to see so many different interactions, often subtle and unexpected, affecting the Arctic ice, the big question that springs to mind is how many of them have a clear connection to the greenhouse gases that we have been pumping into the atmosphere and how many of them are "natural variation," the usual ups and downs of the Arctic.

To put the question more bluntly, do we have to accept that we are to blame for the loss of the sea ice, a question I'd been asking myself ever since I saw that first movie of the changing ice? That question quickly prompts another. Given the speed with that the Arctic is changing, how long will it be before we start to see summers that are free of ice? There are no more vital questions than these for the future of the Arctic seas, the creatures that live there, and the people who make their home around its shores. The only way to tackle them is by turning to computer models of the changing climate.

Based on all the evidence we have, we can be very confident that the Arctic is warming as a result of the rising levels of greenhouse gases in the atmosphere. The most general test came recently when the Climate Research Group at the University of East Anglia looked at all the global climate models that had been used by the Intergovernmental Panel on Climate Change (IPCC), the international body that was set up to provide objective advice on climate change.[8] "Everyone knows that temperatures in the Arctic are rising," Philip Jones, the director of the group, explained to me in 2008, "but the Arctic region had not been looked at on its own until now in computer models of changing global climate."

The Climate Research Group team compared the temperature changes in the Arctic with predictions from computer models that had been run either with or without the extra impact on climate from increasing levels of greenhouse gases. Without the added greenhouse gases, the models predicted that temperatures in the Arctic would stay much the same. Only when the computer models were run with those extra greenhouse gases did the Arctic warm in the way that it has.

Predicting how the sea ice will change and when the Arctic will see ice-free summers is even harder than predicting Arctic temperature changes. Computer models of the sea ice correctly predict that its ice area will decline, but so far they have totally misjudged the speed at which it will happen. That is worrying. Julienne Stroeve of the National Snow and Ice Data Center in Boulder, Colorado, was the first to look at all the climate models and compare them with the real world. The result was shocking, says Stroeve. She summarized her comparison in a graph that has become famous in the sea-ice world.[9] Anytime you sit through a presentation on the future of the Arctic, a version of it will always pop up. The graph shows the average of the predictions from thirteen different computer models of the extent of September sea ice year by year, from 1900 through to 2100, and adds a line showing what has really happened.

In 2006, the real sea-ice cover was "thirty years ahead of the ensemble mean model forecast," Stroeve wrote. And that was before the catastrophic slump in sea-ice cover in 2007 that was repeated in 2008. According to the computer models, the sea ice should not have retreated to the 2007 level until 2055!

"It is happening much faster than expected," says Stroeve, "and now we have 2007 and 2008, it is even more dramatic." So what has happened? Stroeve gives a familiar reply. "The ice has just gotten too thin. We might be moving toward a transition period. Some of the models show that once you have thin ice and you get natural variability that hits on it hard and albedo feedback sets in, you can get a very rapid change."

I went back to Mark Serreze for his view of the state of computer models. As the head of the world's biggest Arctic ice research laboratory, he packs some authority. Serreze is clearly gripped by his work (he tells me he looks at the latest ice data coming into his institute every single day). He punctuates his account of what is happening with a conspiratorial "OK," as though he were drawing you ever deeper into a mystery, which he often is.

Curiously, he began his research career looking at a theory that another ice age was on the way. Back in the late 1980s, some scientists worried about possible global cooling. As Serreze says, "In twenty-five years, it has just changed before my eyes. You think that climate change is a slow process but nowhere in the world can you see it as a more visible effect than in the Arctic."

"We have not quite sorted it out yet," admits Serreze of the gap between model and reality.[10] "If you go back to the climate models and look at what we have seen over the past twenty years, you can see that we are far ahead of schedule in the sea-ice loss. All these climate models are saying that we should be losing ice. There is a consensus that there is a connection to rising greenhouse gases. But none of the models are fast enough in terms of the ice loss that we have seen. Why? One argument is that the climate models are just too insensitive; maybe we haven't modeled the clouds correctly or something like that. The other argument is that climate variability has boosted the trend."

I wonder if any of these different parts of "natural climate variability" that made up the perfect storm, including unusual winds and sunshine, are themselves related to rising greenhouse gas levels and might not be quite as "natural" as they might look. "They could be thought of as natural variability working in concert with a greenhouse forcing," says Serreze. "But what has always puzzled me is that it seems to be a hell of a coincidence. Why is it that all these things are conspiring now? Do we have a complete answer of why we are losing the ice so fast? The short answer is no. I think it's going to be a while before we really have this figured out, and the problem is the changes have been unfolding faster than our ability to understand them. But we're catching up."

The models may have been running well behind the real world but there is a still a lot that can be learned from them. One of the best is the National Center for Atmospheric Research's model called CCSM3 (Community Climate System Model 3). That model does predict catastrophic changes of the kind we are beginning to see, even though the timing is not right. David Lawrence, a climate modeler who works on CCSM3, has been figuring out what it might tell us. "It simulates these types of events where you get big drops of sea ice over periods of five to ten years, which is definitely not on the slow trend line," he says. "Maybe that is what's going to happen in the next decade or so. We may be going into a period where the ice is going to drop substantially and very quickly."

So that crash of 2007 might be the start of a dramatic downward slide. Lawrence explains that they ran the climate model eight times to generate eight different sets of predictions (or "ensemble members"). It is standard practice to repeatedly run the same model with slightly different starting conditions, and then calculate an average prediction (an "ensemble mean"), because each time the model will give a somewhat different result, thanks to the natural chaos of the world.

That mean prediction is rather like the answer you might give to the question: "How long does it take you to commute to work?" You might reply "thirty minutes." What you really mean is that it takes thirty minutes on average but on some days it is less, some days it is more, and there was that day it took two hours when there was big accident. Unlike your commute, which happens every day, the future of the sea ice will happen only once. We can't expect it to be an "average" future. The real future will lie somewhere within the range of the possible.

Within the runs of those eight ensemble members, Lawrence explains that NCAR researchers found nine different "abrupt ice loss events," the earliest of which hit in 2012 and the latest in 2045. "So the model is saying things like this can occur," says Lawrence. "I think if we had run the model one hundred times we may very well have had an event that occurred as early as 2007."

The ups and downs of the run that showed an "abrupt ice event" in 2012 look eerily like the real graph. It is simply running a few years late. The ice disappeared for exactly the reasons that we are seeing now—thinning ice hit by natural variability and the ice–albedo feedback.

That flailing pattern of ups and downs coming out of the computer is unnerving. Uncertainty looks set to rule. In the near future, there may be collapses followed by lulls, and sometimes large recoveries of the ice area, followed by further collapses. Nobody should be surprised if over the coming years and decades there is no simple downward trend in the summer ice minimum. We could have a few years of plateau or recovery. We might all start to think the Arctic is on the mend. We can be certain that the global warming skeptics will be out there shouting "hoax" every year the area of ice does not fall lower than the year before. The computer model shows that a slow and steady decline is not what we should expect at all. There may be lurches downward, pauses, rises, periods of stability, and then further collapses. But what is certain is that the ice is on the way out.

That leads us to the million-dollar question: when? When you ask this question, faces grow grim. Everyone is well aware that we may be witnessing the fastest single environmental catastrophe to hit this planet since the rise of human civilization. There has not been any environmental change on this scale and at this speed anywhere on Earth in the last

2,000 years, especially a change for which humans are to blame. Jim Overland, an oceanographer at the Pacific Marine Environmental Laboratory
in Seattle with a long track record in studying the sea ice, began a talk
to the American Geophysical Union in 2008 by showing a picture of the
retreating ice edge and saying: "I guess I am speaking for the polar bears
this morning. . . ."

So when does he think the ice will vanish? Overland's estimate was
2030. I asked Mark Serreze. "Before 2040," he says. David Barber's view
was, "The earlier end of the range of 2013 to 2030." One way to get the
most out of computer models is to select all those that have made the best
predictions in recent years and have good sea-ice physics built into them.
Muyin Wang of the University of Washington in Seattle, working with
Overland, pulled six models out of the twenty-three now available and
found that if she looked ahead at their predictions, starting from where
the ice has reached (not where models say it should be), then not a lot of
time is left. An ice-free summer may occur as early as 2028 and most probably by 2037.[11]

None of these estimates give us very long, especially compared to the
answer "2100" that was standard only a few years ago. But there is one
computer modeler who predicts a truly scary date: "between 2010 and
2016." And he has good reasons to think he may be right.

Wieslaw Maslowski works at the U.S. Naval Postgraduate School in
Monterey, California. When I spoke to him he was just heading off to the
UN Climate Change Conference in Poznan, Poland, his home country. He
was going along, he explained, at the invitation of former U.S. Vice President Al Gore. Maslowski's models are quite different from the ones we
have talked about so far. His is a "regional Pan-Arctic model for the ocean
and sea ice" rather than a global climate model.

A key point about his model, he explains, is its very high resolution. "Resolution" in a computer model is like that in a digital camera. If
you blow up a digital image enough, you'll soon see the individual pixels:
the camera can't record any detail within a pixel, it can only show you
the average of what was in there. Similarly, climate models break the surface of the planet into "grid squares," although they are also layered in
three dimensions. Each grid square has multiple levels for the different

layers of the atmosphere above and the oceans below the planet's surface. "The global climate models have 100-to-200-kilometer [60-to-120-mile] resolution. The model we are using has a resolution of 9 kilometers [5.6 miles]," explains Maslowski.

High resolution may not matter quite so much when you are modeling the atmosphere where storms, high pressure systems, and so on are large: you only need to look at a weather map to see that they spread over hundreds or even thousands of miles. But down in the ocean and among the ice floes, currents may be only a few miles across, and there are narrow straits in the Arctic where warm and cold waters are flowing in different directions in an area far smaller than a single grid cell in a global climate model.

Maslowski's model also has many layers in the first twenty-five meters (eighty-two feet) of the ocean to give a more realistic fit for the Arctic's shallow seas. With its capacity to capture ocean details that climate models blur, it is no surprise that Maslowski sees ocean effects that he thinks others miss. That is why his model produces a different result, he explains. "Many people do not fully understand the importance of the ocean," he says. "Possibly up to 60 percent or more of the sea-ice melt in the western Arctic region is determined by oceanic heat under the ice." Plenty of oceanographers agree with him, although you won't be surprised to hear that the atmospheric scientists remain cool. We are back at that divide.

At the end of the scientific chase, consensus is not perfect, but we are still left in a bleak place. The summer ice is disappearing. It may all be over within a few decades—or it may be over in a few years. The Arctic will be a new world. That is what is really important. Each winter the seas will freeze over with thin, new ice and each summer that ice will mostly vanish again, except for patches here and there in the most favorable locations, most likely among the Canadian islands. Compare that to the thousands of years that the Arctic has been full of thick, stable multiyear ice, rich with life, each summer. An Arctic that freezes over and melts again each year is a completely different place for the creatures and people that live there now. If change happens that fast, it will be the most unimaginable, wrenching challenge for the Arctic's animals, ecosystems, and people.

There is more to say. Not only are we to blame for the disappearing ice, but we cannot now stop it dead in its tracks by cutting our greenhouse gas emissions. It is too late. Even if a miracle occurred and we suddenly stopped pumping any greenhouse gases into the atmosphere—if industry shut down, if we stopped driving, flying planes, or cutting down forests—the earth would still go on warming for many decades from the greenhouse gases we have put out already. The main greenhouse gas, carbon dioxide, leaves the atmosphere very slowly, and the carbon dioxide we have already added has not yet completed its warming of the earth.

David Beerling, who studies past climates at the University of Sheffield, explains why.[12] "Think of the earth as a bit like your oven. If you set the temperature on an oven, it doesn't just leap to that temperature. The heating element is inefficient and the metal casing takes a long time to warm up. What we have done by adding greenhouse gas in the atmosphere is to very quickly set a new temperature but because the oceans are on average a little over three kilometers deep, there is an enormous amount of energy they can absorb before the land surface and the air temperatures start to rise. An enormous amount of warming will go into the ocean before planetary temperature stops rising."

Of course, greenhouse gas levels are not falling. Only if we took truly dramatic and immediate action to cut greenhouse gases might there be a chance to keep some of the Arctic summer ice, although its area would still be less than it is now. In April 2009, a study at the National Center for Atmospheric Research (NCAR) suggested that if emissions were cut now and greenhouse gas levels stabilized at 450 parts per million (ppm) by the end of the century, the ice might lose only a further quarter of its area.[13] But greenhouse gas levels are already over 380 ppm and on current trends will rise to 750 ppm by 2100, so very urgent worldwide change would be needed. And this prediction is from only one model, while many others show the ice will be gone before 2050.

I asked several of the sea-ice scientists if they thought there was anything else that could be done. "Human beings cannot," says Maslowski, "but a major volcanic eruption might. That could cool the climate and might be a turning point for the ice." A big eruption sends so much dust high into the air that it acts as giant sunshade over the earth.

Ron Lindsay from the Polar Science Center at the University of Washington mentioned volcanoes too: "Something like the Mount Pinatubo eruption [in the Philippines in 1991] might bring back the ice for a few years. If you look at the global mean temperature after Pinatubo, you see a big decrease in temperature, it cools the planet. It recovers within two to three years, though. You would need a long period of volcanism. So don't hold your breath." "I think going to an ice-free state is inevitable," says Serreze. "It is just a question of when. You can't rethicken. It would take too long. It is inevitable now. I hate to say it."

It takes a few moments to absorb the word "inevitable." If there is any upside to follow, it comes in the uniquely powerful message that the Arctic is sending. What stronger sign of climate change could there be than millions of square kilometers of white ice turning to dark water in less than a single lifetime? Is there any more dramatic wake-up call?

If we had the eyes for it, we would already be able to see other terrible changes creeping across the planet. The oceans everywhere are becoming more acidic as carbon dioxide dissolves in their waters. At special risk are any sea creatures with shells or skeletons made of calcium carbonate that the acid can dissolve away. Coral reefs are particularly vulnerable, and fewer than half may survive even current carbon dioxide levels.

We can't see these changes directly, but we have no excuse for inaction. The Arctic has sent a warning that is spelled out in black and white. It is past time to deal with climate change. Otherwise, we will all be using the word "inevitable" a lot more often. And we will all hate saying it.

BORDERS

Chapter Seven

WHO OWNS THE ARCTIC?

When a Russian miniature submarine planted a titanium flag on the seabed beneath the North Pole in August 2007, panic broke out. Was Russia about to seize control of the Arctic Ocean the old-fashioned way? Canada's foreign affairs minister Peter MacKay immediately protested: "This isn't the fifteenth century. You can't go around the world and just plant flags and say, 'We're claiming this territory.'"

Russia's move came as the five nations facing one another across the Arctic grew more assertive. All had announced that they had rights to enormous areas of the Arctic seas—or at least to the oil and mineral wealth lying at its bottom. A month earlier, Canada's Prime Minister Stephen Harper had announced plans for a fleet of ice-strengthened patrol vessels and a new Arctic base. His aggressive line on Canada's Arctic sovereignty was: "We either use it or lose it. And make no mistake, this government intends to use it." A few months before that, Norway's minister of defense announced plans to buy thirty jet fighters because of the nation's "key interests related to the security developments in the High North." And in early 2008, European Union foreign policy chief Javier Solana joined in, claiming that the flag planting had changed "the geostrategic dynamics of the region with potential consequences for international stability and European security interests."

Some Russian newspapers more sensibly described the flag planting as a "stunt" and pointed out that two foreign millionaires, one Swedish and one Australian, had gone along on the trip. As a stunt it was a dramatic success. Nations around the world woke up to the political, economic, and strategic significance of the Arctic, and as the months passed they began issuing new policy directives, rereading legal documents, pouring money into mapping claims to the Arctic, and debating how it should be governed. The Arctic will never be the same again.

The European Union was first off the mark with a strategy document in late 2008,[1] but it was the world's superpower that everyone was waiting for. Just before leaving office in January 2009, U.S. President George Bush issued an Arctic National Security Directive, the first in fifteen years.[2] Its biggest surprise was the emphasis it put on ratifying the UN Convention on the Law of the Sea, despite decades of Senate opposition. Although 156 nations have ratified the convention, including all the other Arctic nations, and although it has been in force since 1994, U.S. conservatives have long argued that it could restrict U.S. sovereignty. Russia soon responded with its own document laying out that it would robustly defend its energy and shipping interests in the Arctic.[3] Other players joined in, too. NATO held a meeting on Arctic security in Iceland at the beginning of January 2009, and the Nordic nations got together to discuss joint air defenses a month later.

The Arctic has not seen so much diplomatic excitement since the end of the Cold War. The one constant in all this activity was the view that the Law of the Sea provided the framework for sorting out Arctic sovereignty. That point had first been stressed in a meeting in Ilulissat in Greenland, hastily arranged by Denmark just a few months after the Russian flag reached the seabed.[4] The meeting's goal was to kill any idea that there would be an aggressive race for the North Pole. The five nations that rim the Arctic seas (United States, Russia, Canada, Norway, and Denmark-Greenland) signed a declaration telling the world that, "an extensive international legal framework applies to the Arctic Ocean," which would provide for the orderly settlement of claims. "The rules are in place," said Stig Moller, Denmark's foreign minister.

Such was the enthusiasm for the "rules" in Ilulissat that everyone forgot to mention that there are several long-running disagreements between these very same Arctic states that the Law of the Sea cannot quickly resolve. Border disputes between America and Canada and between Norway and Russia have continued for decades. There is a less important spat between Greenland and Canada over tiny Hans Island, and the strange case of Svalbard—an archipelago of far northern islands where Norway has sovereignty, but where a curious 1920 treaty guarantees access to many other nations. Here there are fishery disputes with Russia that could begin to involve oil wealth too.

Other disputes were only heightened in this first round of Arctic diplomacy. The United States reasserted its long-held position that the Northwest Passage, which winds among Canada's islands, is an "international strait," not an "internal waterway." Other nation's ships thus do not need Canada's permission to pass through it. That view, supported by Europe, has been enraging Canadians ever since 1969 when the United States sent the giant ice-breaking oil tanker *Manhattan* through the passage. The United States also regards key straits on the northern route around Siberia as "for international navigation," a view which Russia does not accept.

Those who were pushing for a comprehensive new treaty to bring peace and stability to the Arctic were also disappointed. In October 2008, the European parliament had passed a resolution demanding the start of international negotiations;[5] earlier, the World Wildlife Fund (WWF) had published a report laying out the logic for a new treaty.[6] One much-cited model was the Antarctic Treaty, which has successfully frozen territorial claims, kept out industry, mining, and the military, and preserved Antarctica for peaceful scientific cooperation. But the U.S. directive bluntly stated: "An Arctic treaty of broad scope—along the lines of the Antarctic Treaty—is not appropriate or necessary."

Also sidelined were calls for the Arctic Council to play a bigger part in governing the Arctic. The Arctic Council is the region's most important high-level forum, with representation from the eight Arctic states (the five Arctic Ocean states, plus Iceland, Sweden, and Finland) along with permanent participation from the Inuit Circumpolar Council and five other indigenous groups. A range of nongovernmental organizations have permanent observer status, as do representatives from France, Germany, Poland, Spain, the Netherlands, and the United Kingdom. The council has no authority to make laws or set regulations, but it has been able to steer the priorities of the Arctic nations and issue authoritative reports (on pollution and Arctic climate change, for example) which have driven action by other bodies. Although the council lacks power, it has

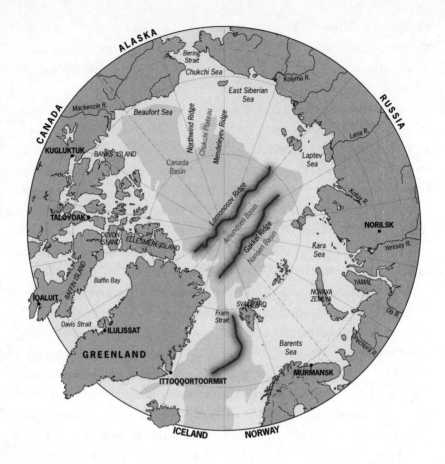

enormous influence, which is why China, Korea, Italy, and the European Union are lining up to gain observer status. The U.S. directive said it was not to be "transformed into a formal international organization." The message is clear: it is to the Law of the Sea, rather than big new treaties, that we must turn to look at the future governance of the Arctic, at least in the next decade when change is at its most rapid.

The first thing that the Law of the Sea can do is to allow claims for a bigger slice of the Arctic seas to be assessed in an orderly way. Article 76 of the Law insists that a claim to sovereign rights over an "extended continental shelf" must be driven by geological data and not flags. Claims go for examination by the twenty-one experts that make up the Commission on the Limits of the Continental Shelf at the United Nations headquarters in New York, and they need to be submitted quickly. Each nation has only ten years from the date it ratifies the convention.

Russia put in its claim promptly in 2001. The map of the claim (which is public, although the data behind it are not) shows great ambitions: Russia would have gained rights over half the Arctic, running right up to the North Pole. The commission cannot reject claims, but it can send them back. That is what it did, saying "more geological evidence needed." Russia has been busy sending out Arctic survey ships ever since.

The Law established that all coastal nations have a right to an exclusive economic zone (EEZ) that extends 200 nautical miles from the "baseline" of their shores. That is undisputed, and thanks to the many islands in the Arctic, immediately takes huge chunks out of the seas. Further claims can be made out to an absolute maximum of *either* 350 nautical miles from the baseline *or* 100 nautical miles from the 2,500-meter (1.5-mile) depth contour line, whichever of these distances is the greater. The Arctic is ringed by shallow seas (a little over half the entire Arctic seas are shelf) so there are many places where the waters pass well beyond the 350-nautical-mile limit before their depth falls to 2,500 meters; most are off Russia where the shelf can extend 500 nautical miles from land.

Several other strict criteria have to be met in order to claim seabeds within those final limits. States must show that the sea bottom they claim is a "natural prolongation" of their own land, that is, it is not just joined physically to the land but sharing the same geological history. If that condition is satisfied, nations can claim out to the "edge" of the shallow shelves (provided they don't exceed the limits), where the continent ends and the abyss of the deep seas begins.

To complicate matters further, there are two different ways to define the "edge." The first makes immediate intuitive sense: it is sixty nautical miles past the "foot" of the slope, which is where the steep drop at the shelf edge levels off after it reaches the deeper ocean floor. The second is more puzzling. The edge can be placed where the thickness of sediment on the seabed is "at least 1 percent" of the distance back to the foot of the slope. That means the edge could be placed a hundred miles out from the foot of the slope if the sediment there were one mile thick.

At this point I needed help and I turned to Ron Macnab, a Canadian who is the father figure of Arctic mapping. Macnab began his career as a naval officer and used to spend "hours standing watch on the bridge, staring out to the ocean wondering what in the world is going on, in and under it." Working with the Geological Survey of Canada, he helped create the International Bathymetric Chart of the Arctic Ocean. He still sports a neat naval beard and moustache, but these days he is "retired" which means he runs around chairing meetings on the Law of the Sea and Arctic geology.

The rule about sediment thickness, he says, is called the "Irish formula," after the three Irishmen who developed it. Sediment slides off the edge of the continental shelves and builds up thick layers in the deeper sea alongside. These thick sedimentary basins are exactly the kind of place where oil and gas is found. "The rule provides a framework for ensuring the coastal states will have a worthwhile piece of the sedimentary basin in which they could drill for oil," he explains. The thickness of the sediment gradually tapers off farther out to sea and the formula provides a cutoff within which coastal states have rights to a potential source of wealth.

To make a claim under the Law of the Sea, explains Macnab, the essential data you need are the depth of the seabed and the depth and distribution of sedimentary material below it. The depth data are obtained using sonar to ping the bottom and record the returning echo. To look at the sediment beneath requires seismic profiling, which involves making a series of very loud noises that penetrate the seabed and echo back from the layers of sediment beneath to an array of microphones.

"In principle, the rules are quite straightforward," says Macnab, "but in fact, there is a lot of ambiguity in it, starting with how you define the foot of the slope. The convention was a consensus document which meant that it left a lot of language that was quite vague so nobody could feel pinned down. Article 76 is only a little over six hundred words long, but the Commission on the Limits of the Continental Shelf labored mightily to produce about a hundred and thirty pages of explanation about what the words mean. I can tell you from personal experience that an investigator can look at a profile one day and say this is where the foot of the slope should be and then come back the next day and think maybe it should be somewhere else. It is a messy proposition."

Collecting data in the Arctic is especially difficult. "Bathymetry and seismic rely on the transmission of sound data through water, but if your ship is crashing through ice you are degrading the quality of your measurements," says Macnab. "Often you can't go where you really want to collect data, you have to make detours, you have to run zigzag tracks around ice instead of a nice straight profile."

So to get started, claimants need to overcome these difficulties and prove that they have accurately located the foot of the slope, the sediment thickness, and the location of the 2,500-meter line. With only 6 percent of the Arctic already mapped with advanced sonar techniques, there is a lot of basic bathymetry to do. That is the major part of the job where areas of seafloor are accepted to be a "natural prolongation" of the land. Much of the huge area of shallow sea that abuts Siberia and Alaska, and the narrower zones around Greenland and Canada, are obviously prolongations of the land but must still be accurately mapped to fix the borders. Along the way there can be some surprises.

Larry Mayer, director of the Center for Coastal and Ocean Mapping at the University of New Hampshire, is the United States' number-one undersea mapmaker. He has created dramatic three-dimensional imagery of the seabed, using multibeam sonar which sends out a fan of sound to scan the ocean floor. The difference between the old-fashioned single-beam sonar, which just produces a series of dots, and multibeam is "like going from an outline of a face to a color photograph," says Mayer. The U.S. Arctic mapping effort has repeatedly taken him out to the Chukchi Plateau and Northwind Ridge—extensions of the continental shelf off Alaska—on the icebreaker USCGC *Healy*. Out there, Mayer scored a territorial win for the United States when he found that the foot of the shelf was over a hundred miles farther north than previously thought. That will give the United States rights to thousands of extra square miles of Arctic seabed, once the Law of the Sea has been ratified.

There were plenty of other wonderful new sights too. One highlight was discovering a new undersea mountain, now named the Healy Seamount, almost two miles high and nine miles wide, out near the Chukchi Plateau. Elsewhere, the multibeam sonar provides stunning imagery of seabed features. There are long scour lines that have been made by giant icebergs dragging across the bottom, grooves cut by ancient ice sheets, and a couple of features that I had never heard of before including "pock marks," scores of little craters where gas has bubbled up, perhaps from methane hydrates breaking down. Then there are "mud volcanoes." Pressurized gas bursts through a muddy seafloor, throwing up a huge volcano-shaped mud structure that slowly collapses after the eruption is over.

Arctic geologists run into bigger problems and potential conflicts on the chains of ridges (the Gakkel Ridge, the Lomonosov Ridge, and the Alpha Mendeleyev Ridge) that lead out across the deeper Arctic. It is not so obvious that they are natural prolongations of adjoining land, but if they are, there's a chance for a claim that runs far out across the Arctic.

The Gakkel Ridge is the easiest to deal with. It is the northern end of the better-known Mid-Atlantic Ridge, a great zipper in the earth's crust that runs all the way down past Iceland to beyond the tip of Africa. Here convection currents deep within the earth bring fresh, hot magma up to the surface as the Eurasian and North American continents drift apart. Article 76 rules are clear for the Gakkel Ridge. This spreading ridge is "oceanic." No one can claim it.

The Lomonosov Ridge is quite different and seems to be a huge fragment of continental crust left behind as the continents moved apart. It has turned and subsided over millions of years and now forms an underwater range of mountains, covered in layers of sediment, which stretches 1,100 miles right across the Arctic from Russia to where Denmark's Greenland meets Canada's Ellesmere Island.

The Lomonosov is a candidate to be a "natural prolongation" of land belonging to Russia, Greenland, and Canada depending on which end of the ridge you start from. There are difficulties with this: when the commission looked at the first Russian claim to the Lomonosov, the rumor is that it saw "morphological breaks" in the ridge, that is, dips in the ridge suggesting it was not linked to the land. "This raised the alarm in Canada and Denmark, too, because there are similar morphologies at the landward ends of the Alpha and Lomonosov ridges," says Macnab. "What's needed is to determine that there is geological continuity from the continent through to the ridge underneath that morphological break," he explains. Researchers need to dive beneath the seabed and look inside the ridge to plumb its deeper structure. That involves some very loud bangs.

The Danes and Canadians joined forces in Project LORITA to work on their end of the Lomonosov. Landing on the ice by plane and helicopter they laid out 150 hydrophones at 1.3-kilometer (0.8-mile) intervals over a series of 200-kilometer (124-mile) lines. Under the ice they hung eleven "shots" per line, each holding 450–900 pounds of Pentolite high explosive that can send out sound waves penetrating 40 kilometers (25 miles) deep into the ridge below. Each shot could have demolished an entire office block. Mapping work continues and will determine how strong a claim the two nations have at their end of the ridge. The Russians are busy at the other end.

Further around the ocean, we run into the Alpha Mendeleyev Ridge, which is mysterious, little explored, and of disputed origin. When the Russians went ahead and claimed the ridge back in 2001, the United States objected immediately. The ridge, they said, is a "hot spot," where plumes of magma from deep within the earth have intermittently produced volcanoes on the overlying crust above. The U.S. statement concludes that the ridge is "a volcanic feature of oceanic origin. It is not part of any State's continental shelf."

Russian scientists who used a whole battery of geological techniques to explore the ridge on a 2005 expedition do not agree. They say the evidence will show that it is a submerged block of continental crust and very much a natural extension of Russia. Canadians are reaching a similar conclusion from their end.

Mapping is continuing, submissions to the commission are not yet in, and any judgments are five to ten years in the future. Still it is possible to begin to work out what a future map of the Arctic might look like by listening to the scientific reports that the geologists presented at the International Geological Congress held in Oslo in August 2008. At a meeting like this, it is all science and no political intrigue. Macnab says that is the best part of his job. "Our governments and our authorities clash over some of these issues and these may lead to international disputes. But we develop real friendships," he says.

Looking over the abstracts of their research, they certainly don't seem to contain politically motivated exaggeration, just rather a lot of data.[7] Those data show that every nation may have a reasonable claim on the ridges crossing the Arctic. "If those observations are believable," says Macnab, "and I have no reasons to consider that they wouldn't be, then to me that represents that we are looking at genuine prolongations of the land masses." But the members of the commission may not agree with that, he cautions.

If all prove geologically sound, then Danish, Canadian, and Russian claims will overlap in the middle of the Arctic; American, Canadian, and Russian claims in the east and Norwegian and Russian claims in the west. Everyone will agree that there are two "holes," areas to the left and right of the North Pole, which no one can reasonably claim as they are clearly beyond both those maximum limits. So the question of who owns the Arctic becomes one of how to settle geologically sound overlapping claims.

"The commission itself will not act as arbiter here," says Macnab. "If there is a situation where the two countries can't agree, the commission will hand it back to them and say, 'You have to deal with it. It's your problem. We are only looking at the technical aspect of your submissions, not the political ones.'"

Before guessing as to where that might lead and what the future political divisions of the Arctic will look like, it is worth asking what prize is really at stake.

The prize is not big or immediate. Most of the oil and gas basins of the Arctic lie in the 200-nautical-mile exclusive economic zones. Further out there will be more hydrocarbons, but easier prospects will be available for at least fifty years. Deep in the Arctic there are likely to be gigantic methane deposits. Mining methane hydrates is still at the experimental stage. Mining subsea methane in remote and deep basins covered by sea-

WHO OWNS THE ARCTIC? 113

sonal ice is a very distant possibility. Fish are another resource, but rights
to them come only if they live on the sea bottom. Within the EEZ there are
exclusive rights to fish within the sea, but on the "extended continental
shelf" rights are more limited. The bottom of a cold, deep ice-covered sea
is not full of fish.

That means that rights to the extended shelf really aren't of any immediate value; they grant no control over shipping or air traffic and certainly
aren't going to allow anyone to set up a customs post at the North Pole. If
there is a rush, it is not because there is something of immediate value to
fight about but for a much more mundane reason: the deadline for filing
claims is approaching. Russia and Norway already have claims in place,
and Norway's claim in the Barents and Norwegian seas was approved in
March 2009, although that does not resolve border disputes. Canada has
until 2013 and Denmark until 2014. The United States has yet to ratify
the United Nations Convention on the Law of the Sea (UNCLOS), so it has
ten years after it does. It is worth claiming because if a nation doesn't, any
benefits, however unlikely, will be lost. Back in 1867 when America bought
Alaska from Russia for $7 million, critics attacked the waste of money on
land that contained "nothing of value but furbearing animals." No one
thinks that money was wasted now.

The tight deadlines do have another consequence. They apply to all
UNCLOS claims, everywhere in the world, and right now there is a global
rush. The ten-year deadline for nations that ratified the convention early
on was set at May 13, 2009, and submissions from more than fifty nations,
from Ireland to China, have arrived at the United Nations. One estimate
is that it will take the commission thirty-five years to clear the queue.

For lessons on how overlapping claims might be settled, ongoing border disputes are instructive. They are also important in their own right,
especially as two of them are in oil-rich areas. The first of those is the
United States versus Canada. The land border between Alaska (United
States) and the Yukon (Canada) for the most part runs straight along the
141st line of longitude, a line that if you follow north will run straight
to the pole. The demarcation line was agreed in 1823 in negotiations in
Moscow between Great Britain and Imperial Russia, written in French,
and passed on when Alaska was bought by the United States and Canada

became independent. What happens when that straight-line land border reaches the sea? The Canadians say the line should simply continue straight on and neatly divide the coastal waters. The Americans say no: that line stops at the shore and the coastal waters are divided by a principle of equidistance from the adjoining land masses, which means the line pretty much goes at right angles to the shore, not straight on toward the pole. Of course, each nation's preferred method gives it more of the coastal waters.

That great chunk of ocean under dispute is close to oil-rich areas of Prudhoe Bay and to the Mackenzie Valley and is likely to be valuable. The Canadian government has issued more than a dozen exploration permits to oil companies nearby in the Beaufort Sea. Both nations want to explore the disputed zone, and the higher the price of oil and gas, the lower the chance anyone will back down. I asked Rob Huebert, an outspoken commentator on Arctic security issues and associate director of the Centre for Military and Strategic Studies at the University of Calgary, if it would be settled soon.

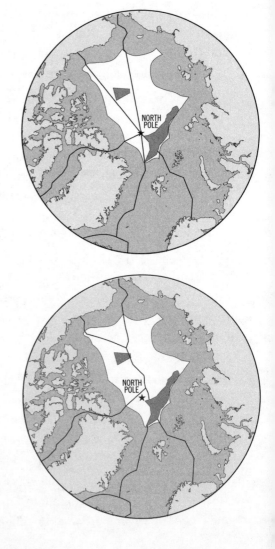

The longer the issue is left, the more people worry about what they might lose and the more dangerous it becomes, Huebert explained. "We should have done it five years ago. We should have had a joint management scheme for oil and gas, continued to agree to disagree about where the official boundary is, and made sure that any development in the disputed zone followed the highest environmental protection standards."

Over the other side of the Arctic, it is Russia versus Norway. Here, it is Russia that backs a border that runs along a line toward the pole from the land boundary. Norway backs the equidistance method. The disputed zone is large and, unsurprisingly, each nation backs the method that gives it the bigger area. The dispute is holding back oil exploration, and both Russia and Norway appear keen to solve it. Every year there is a rumor of an imminent breakthrough.

If the whole Arctic turns out to be covered with overlapping successful claims, as looks increasingly possible, then the "sector-line principle" and the "equidistance principle" provide two rational ways to resolve the borders of these claims. Macnab has charted these alternatives, and they have very different consequences.

Ironically, Canada would gain far more of the Arctic overall by supporting the equidistance principle (shown at the bottom), although it would lose if it applied it specifically in its border dispute with America. "There are other geometric options," Macnab says. "The point is that the Arctic coastal states have to be prepared for bilateral and multilateral negotiation when the time comes." Nations can take a disputed boundary to the International Court of Justice in the Hague or the United Nations International Tribunal on the Law of the Sea in Hamburg, but more often negotiation is preferred to the uncertainties of an outside court settlement.

Even so, we can guess who is going to own one very important bit of it. The Danes and Greenlanders are likely to end up with the North Pole. Of course, if nations opt for a sector-line method (shown at the top), then all claims will meet at the pole, but otherwise it is just too close to Greenland for it to easily fall within anyone else's claim. We probably won't have that confirmed for at least a decade. Long before then, the nations surrounding the Arctic will have to confront many other issues of governance raised by the vanishing ice.

* * *

There is a lot more to looking after the Arctic as it grows more accessible than settling claims to the sea bottom. Claims to the "extended continental shelf" only provide rights over the seabed, not the "water column," so the huge areas of the Arctic seas beyond the 200-nautical-mile EEZ remain open to all.[8] Even within the EEZ, powers are restricted. A nation has sovereign rights "for the purpose of exploring and exploiting, conserving, and managing the natural resources" of both waters and seabed, but it cannot control ships passing through these waters nor impose regulations—to prevent pollution or the dumping of trash, for example—that exceed international norms. Nations only have complete sovereignty over their "territorial seas" (the first twelve nautical miles of their seas), and even there ships have rights of "innocent passage."

If there are no international rules to stop them, ships that aren't really suitable for the Arctic could begin to travel there and risk oil spills as the ice retreats. Fishing boats can chase fish as they move into the newly warming Arctic if they aren't stopped from doing so, and boats can easily carry alien species into the warming waters. Fishing boats and ships are very mobile, and fish, whales, birds, and drifting oil spills don't recognize national boundaries, so it is not effective for each part of the Arctic to make up its own regulations. Nor can bodies which are responsible for just one sector (shipping or fishing, for example) build the best set of rules to protect the Arctic if each tackles its problems independently. Stresses from different causes (pollution plus overfishing, for example) add up, so the only effective way to look after the Arctic is by what is now known as "ecosystem-based management" that looks at all the impacts from different causes as a whole.

International cooperation is going to be needed, and governments are again turning to the Law of the Sea to look for the framework within which new rules can develop. Hidden deep within the Law of the Sea are articles that can be used to help, and lawyers are now looking at them closely to try to avoid having to build whole new treaties from scratch. Article 234 on "Ice-Covered Areas" is important. A mere 104 words long, the article was inserted at the insistence of Canada, with its perennial worry that international shipping would come plowing through its Arctic straits, ignoring Canada's view that they are internal waterways. Article 234 allows states to apply rules on pollution that are stricter than international standards within their exclusive economic zones if they are "ice-covered" and if pollution "could cause major harm." In 2009, Canada took advantage of this article with an amendment to its Arctic Waters Pollution Prevention Act that set tough rules out to the 200-nautical-mile limit.

Article 211(6) on "Pollution from Vessels" provides opportunities to protect defined areas that have special "oceanographical and ecological" conditions, after consultations through "the competent international organization." The International Maritime Organization (IMO), a United Nations body based in London, is one such organization.[9] Annexes to its International Convention for the Prevention of Marine Pollution from Ships (MARPOL) allow "special areas" to be protected, which can include entire seas. (The Mediterranean and Baltic seas have special protection against oil spills, for example.) Critical habitats can be identified as "Particularly Sensitive Sea Areas." Once an area is approved (as the seas around the Galapagos Islands and the Great Barrier Reef have been), maritime activities can be controlled and ships rerouted.

Article 211(6) only specifies how ships must behave (regarding their "discharges or navigational practices") and cannot require them "to observe design, construction, manning, or equipment standards other than generally accepted rules and standards." But this exclusion, which could allow ships and crews unfit for the Arctic heading for its waters, could be overcome by action on another IMO code. It has a voluntary Polar Code (the IMO Guidelines for Ships Operating in Arctic Ice-Covered Waters) which could be strengthened and made mandatory under its International Convention for the Safety of Life at Sea (SOLAS), which all IMO signatories must obey.[10]

The Law of the Sea also provides considerable powers to the ports at which shipping calls. Article 219 allows port states to inspect and detain unseaworthy ships; Article 218 allows them to investigate ships that may have breached MARPOL rules even if the breaches occurred outside their own economic zone.

Fish that move back and forth between the economic zones of two

nations or are highly migratory are covered in two well-known Articles
(63 and 64). And it could even be possible to invoke Articles 122 and 123,
which instruct states bordering a "semi-enclosed sea" to coordinate its
management and conservation. Many other such seas, including the Mediterranean, have done so under the auspices of the United Nations Environment Program.

It is easy to see why many lawyers believe that the Law of the Sea provides the "Constitution for the Oceans," from which wider protection of
the Arctic could flow. Already there is a buzz of activity, although conclusions are still a while away. One group of experts, calling itself Governance
in a Rapidly Changing Arctic, is hoping to be ready with recommenda-
tions by fall 2010, according to testimony given by one of its members,
Robert Corell, chair of the Arctic Climate Impact Assessment, to the U.S.
House of Representatives in March 2009.[11]

Jacqueline McGlade, director of the European Environment Agency,[12]
is one of the group's leading members, and I asked her where this initia-
tive might go. She explained that a first step is to create a compendium
of all the various treaties which might relate to the Arctic. "We've got our
eyes on all the different legally binding obligations that the various organizations in the countries have," she said.

The list of potentially useful conventions quickly grows: it would include the OSPAR Convention on the North-East Atlantic Marine Environment, the Convention on Biological Diversity, the Basel Convention on
Hazardous Wastes, and the Convention on the Conservation of Migratory
Species on top of the IMO conventions on safety of life at sea and marine
pollution. At the same time, the group is drawing together all the governance issues for the Arctic by talking to every group with an interest, from
national states to indigenous peoples.

"At this stage we are minded to push for something to sit within the
Law of the Sea," says McGlade. "The Arctic Council would continue to
provide the information base and could shine the spotlight on noncom-
pliance." If all the problems that are raised can be met by the conventions
that can be put together, there will still be a need for an agreement to enforce them and a way of providing a presence in the Arctic. "I won't say
a police force, but certainly a presence in the Arctic to protect the pub-

lic good," says McGlade. "That is most likely to come from ships and ice-breakers of the Arctic coastal states."

There have been suggestions that a new agency of the United Nations should coordinate action in the Arctic, but any move toward more centralized governance for the Arctic has to avoid offending the Arctic states. Mead Treadwell is chair of the U.S. Arctic Research Commission, which advises the president and Congress on Arctic research, and has lived in Alaska for more than thirty years. "In Alaska we've worked very hard to get self-determination in the Arctic, including economic self-determination with the Prudhoe Bay oil field," he explains. "The regions in the Arctic, whether nation state or regional government, all believe that we can manage this area responsibly and we've got cooperative mechanisms to do it, with the Arctic Council and the Northern Forum [a forum for circumpolar regional governments]," he explains. "Denmark is trying to give stronger local governance to Greenland and in Nunavut everybody points with pride to their strengthened local government. Another branch of world government laid on top of it will not work with the Arctic peoples. What will work is if Arctic peoples get together and say, how do we protect ourselves from rust buckets sailing through the Arctic? We will go together to the IMO."

There are inevitably suspicions of nations from outside the Arctic taking a big interest in what happens there. States that have an Arctic coastline are unlikely to accept constraints on what they can do from organizations that include non-Arctic members. When the European Union put out its strategy document on the Arctic, a joke going the rounds in Canada was that the best response would be a strategy document on Canada's plans for the Mediterranean. The EU application for observer status at the Arctic Council (along with applications from China, Japan, and Italy) was not approved at the council's biannual ministerial meeting held at Tromsø in April 2009 and must wait another two years for the next such top-level meeting. The Premier of Nunavut, Eva Aariak, had publicly objected to EU membership, given the Union's plans to ban the import of seal products, but Canada and Russia also had concerns about a stronger EU presence in the Arctic.[13]

No EU nation faces the Arctic seas (Greenland left the EU in 1985 over its fishery rules), and the major strategic concern of Europe is to ally itself with Norway (which twice rejected EU membership) and its Arctic oil and gas supplies. Undersea pipelines from Norway's gas fields already supply one third of the natural gas used in the United Kingdom, Germany, and France. Just before the EU Arctic strategy was announced in November 2008, Norway's Prime Minister Jens Stoltenberg was invited to visit Brussels where he told reporters, "You can count on Norway when it comes to energy supplies to Europe."

Even within the Arctic nations, not all are equal. When the Arctic nations met in Ilulissat in 2008, only five were invited. Two Arctic nations, Sweden and Finland, were excluded because they do not have Arctic coastlines. Iceland, which lies just below the Arctic Circle, but would play a big role if Arctic shipping routes developed, was not invited. The Ilulissat meeting was a reminder that the Arctic coastal states see the Arctic Ocean as their own lake. That immediately angered indigenous peoples around the Arctic who were not invited. They held their own meeting in Canada in November 2008 and registered their complaint that governments were entering into Arctic sovereignty discussions without their involvement; later they described all the Arctic nations as "outsiders."[14]

With all this jostling and scuffling, it is easy to see why the Arctic is emerging as a very busy region for diplomats, lawyers, and activists. There are organizations that represent national interests, subnational interests (the Northern Forum), international interests, transnational interests (like the Inuit Circumpolar Council), and sectorial interests (fishing and shipping, for example). All are busy now. Inuit are making declarations of their right to be consulted on Arctic developments, the International Association of Classification Societies is trying to unify rules for the "winterization" of polar shipping, the International Standards Organization is busy setting a new code for Arctic oil rig safety; everywhere you turn there is hectic activity. No one yet knows where this ferment will lead. Arctic nations and all these other groups may cooperate, and quarrel, over long lists of things that each group feels are urgent. Perhaps a bigger set of rules and structures may be successfully put in place within the overall constitution provided by the Law of the Sea. Or maybe the five nations that front the Arctic will try to slam the door on everyone else. We may see cooperation, conflict, or a patchwork of partial solutions.

The worst prospect is that the Arctic may simply outrun any attempt to govern it. In the United States, it takes an estimated ten to fifteen years to win the budget for a new icebreaker and then design and build it. In a Korean yard, a new commercial icebreaking oil tanker can be ready in three months. If the ice vanishes in five or ten years, not a hundred, the government of the Arctic and the protection of its environment and people could easily slip out of control while there is still just a babble of competing voices.

Chapter Eight

THE STRANGE CASE
OF SVALBARD

Flying into the town of Longyearbyen (latitude 78°13' north) in the islands of Svalbard, I am just 800 miles from the North Pole. As the plane starts to descend, the view is of snowy mountains, ice caps, glaciers, fjords, and huge expanses of brown tundra. But there's something I didn't expect at all. A few minutes before landing, out on a stretch of tundra by the sea, the plane flies past row after row of yellow apartment blocks, set around a cluster of enormous industrial buildings. For a moment I wonder if I'm hallucinating: who would build a high-rise housing complex up here in the High Arctic? A minute later the scene shifts again and now I'm in a science fiction movie. Below me, on an isolated dull-brown plateau streaked with snow, are white radar domes; I just have time to count seven or eight large ones, but smaller ones are scattered among them.

Then I land at Longyearbyen. A giant factory chimney is the first thing to catch my eye on the bus into town; then I see ramshackle industrial buildings and old docks and lines of wooden towers cutting across the hills. I've taken a two-hour flight from the very top of Norway and across the wild sea to visit an Arctic wilderness, but instead I seem to have arrived in a rustbelt industrial heritage park.

That first impression turns out to be correct. The docks were used to export coal brought from mines scattered around the area by a network of ropeways; the wooden towers are all that remain of them. The valley sides behind the town are still scarred by old workings and enormous heaps of spoil. Longyearbyen grew out of a nineteenth-century Norwegian mining town, although mine tours for tourists are now bigger business than the one small local mine. The yellow apartment buildings I saw from the air were real, too: they belong to a large and active coal-mining settlement called Barentsburg, twenty miles away. The settlement is Russian and the architecture is Soviet, although most of the miners come from Ukraine. The Norwegians have another big mine at Svea, farther to the south.

The radar domes are part of a satellite tracking station.[1] A little farther out of town are large radar dishes used to study the aurora.[2] Less easy to spot is a simple concrete structure just outside town that marks the entrance to an old mine which has been converted into a seed vault where samples of the world's most important crop species are stored against global catastrophe. All are symbols of another Svalbard oddity: the islands are overrun by scientists. Four hours farther to the north by ship, the scientists even have a town all to themselves. A former mining community called Ny Alesund is now entirely inhabited by scientists from around the world (plus, I was to find, a colony of Arctic foxes that live right under the scientists' cafeteria). Here, among other odd buildings, you'll find a Chinese research station with an entrance flanked by traditional stone lions.

There is an excess of wildlife and wilderness here, too. Longyearbyen may be a mining town but the wilderness penetrates into its center. Everywhere there are the bright white flashes of the snow bunting's erratic flight, Arctic terns mob you aggressively if you step off the road into their territory, and purple sandpipers run around like little clockwork toys, chased by tiny chicks desperate to keep up with their parents. Even reindeer wander about, searching among the houses and old mine towers for patches of green-brown grass to eat. They are probably safer here than on the tundra: signs warn you not to stray without a rifle to protect yourself from polar bears. The rule is that you must unload your gun when you reenter the town.

I have come here partly to hike in the mountains and travel around the coast to see some of the prodigious wildlife of the Arctic close up, and partly to explore Svalbard's often violent history and learn how it has led to the strange system under which Svalbard is now governed. It might provide a lesson in how competition and conflict in the Arctic can turn into cooperation—or that at least is the good-news story I was told by researchers who work at Ny Alesund before setting off.

A good place for a short course in the region's past and present is Longyearbyen's new little museum, built alongside the University Centre in a long low building with the fjord and mountains behind it. You have to observe a local custom of taking off your (usually very muddy) shoes to enter. Inside, a maze of wooden floors leads you around exhibits set inside a large hall. Here the view is subtly Norwegian; although I didn't know that the first time I visited. Later, at the Russian settlements, I was to hear a slightly different view. And once back home, I was to find that several European nations had yet another opinion of Svalbard and the international treaty that granted a peculiar form of sovereignty to Norway. But let's begin at the beginning.[3] I am glad that the museum ignores Henry Hudson, the English explorer who came to Svalbard in 1603 and unwittingly unleashed a wave of death and destruction, which soon spread across the entire Arctic and eventually down to the Antarctic too.[4] The natural world has never forgotten him.

Henry Hudson did not discover the islands, but he arrived here just seven years after William Barents, the Dutch navigator who did. (Although Russian historians say that's all wrong and Russians were here first.)[5] He then sailed back to London and broadcast tales of the enormous numbers of fat bowhead whales swimming lazily in the islands' untouched waters. Within a few years, scores of whaling boats from all over Europe arrived. Hudson himself knew nothing of the slaughter he had set in motion; he was already dead. Just four years after visiting Svalbard his crew mutinied and abandoned him, his son, and some loyal shipmates in an open boat in the Canadian bay that still bears his name. He was never seen again. (Unless, that is, you believe an old Inuit tale of a hunting party that found a boat full of dead white men and a living boy who they took away.)

The fat bowhead whales did not stand a chance against the European onslaught. The whalers killed them, stripped them of their blubber, rendered it to oil, and returned home as quickly as they could. They were there to get rich quick, and before long they were robbing one another as well as killing the whales.

At first the rivalry was between the British and the Dutch, but they soon agreed it would be more profitable to divide the area up and keep everyone else out. So the Dutch took the area north of the Magdalenfjorden, a stunningly beautiful fjord surrounded by steep mountains, and built a whaling station they called Smeerenburg. Blubbertown, as its name translates, was briefly home to 200 people who worked at giant cauldrons of whale blubber. The English took the area to the south. The French tried to muscle in. On their first attempt to plant their flag in 1613, they were set upon by six armed English vessels and sent packing.

In less than fifty years, the whales were wiped out in inshore waters. Whaling vessels followed them offshore, and once the whalers had learned how to strip whales of their blubber at sea, they took their ships and their skills to the North American Arctic and on to the Antarctic, wiping out the world's whales everywhere they went. (Between 1610 and 1915, 39,251 voyages were made in the Arctic to hunt bowhead whales; the peak decade of the 1720s saw 3,001 voyages.)[6] The Longyearbyen museum records how in 1610 an English whaling captain wrote that he found it hard to sail into the nearby Isfjord becasue it was so crowded with whales.[7] Standing by the same bay now, I can see only cold, gray water. Scientists in Svalbard told me that they think there are still eight bowhead whales in the area. They can't be sure if they are the last descendants of Svalbard's whales or a few strays that have wandered over from the Canadian Arctic.

Although the whales were gone, Svalbard's bloody seventeenth century was not at an end. It remained a whaling base and in 1693, the French were back again to chase the Dutch to Sorgfjord and confront them in what remains the most northerly naval battle ever fought.[8] I passed by Sorgfjord in August, just a few days after the 315th anniversary of the battle; it is a lonely spot, truly at the end of the world. The entrance to the fjord is quite hard to pick out as it is partly hidden by a curving promontory of low land. Farther back, the promontory melts into a plain which leads in among the mountains and back to the enormous ice cap that covers the interior of Spitsbergen. It is not a place to linger even in

August, as the northern pack ice is drifting only a few miles offshore. If the wind changes and the ice edge advances, you may be trapped until summer comes around again.

The Dutch had set up a battery of guns on the promontory, with forty whaling ships arranged in a crescent formation just beyond the narrow entrance. The French had just two frigates, but a daring commander and heavier cannon. Five hours later, the French ships had each fired around 1,600 rounds from their cannon, captured thirteen ships, and killed a great many Dutch. Some were buried here. A lonely cross stands on a low hill nearby, silhouetted against the snow. It is a long way for two nations to come to kill one another over whales.

I dwell on this history because in Svalbard the past lingers on into the present, both in its political arrangements and in a way that directly confronts your senses. In Svalbard it is cold and dry, and the ground is frozen just beneath the surface. For half the year, it is dark. Decay slows almost to a halt. Where the whalers operated in the seventeenth century, hundreds of graves of men and many more bones of the animals they hunted still remain. Men died in the quarrels between nations, of scurvy, and by the whales they hunted. Some still lie in their graves remarkably unchanged. Archaeologists have found them still clothed in well-preserved jackets and knee-length trousers with little knitted hats on their skulls.[9] They have learned much about the clothing fashions of the early seventeenth century.

There are piles of whale bones too, lying almost engulfed in moss and still gradually giving up their form to the tundra even though they died 400 years ago. The larger bones provide a little shelter for wildflowers that nestle alongside them and pick out their fading shape in spots of bright color. Elsewhere, well above the shore, there are scatterings of ancient walrus bones. It is not too hard to imagine the grim scene. An exhibit back at the museum in Longyearbyen explains that walrus aren't easy to kill because of their thick skins and even thicker skulls. A blow from a wooden club that will finish off a seal merely makes a walrus angry. Inhuman ingenuity devised long lances to stab the walrus through a weak point in the neck instead. To kill as many as possible, the trick was to attack a colony of walrus from the water's edge. Unable to flee to the sea across the mounting piles of the dead, the panicking herd would clumsily stampede inland where they could be completely wiped out.

Walrus were hunted especially by the Pomors of the White Sea, Russians who stayed on in Svalbard long after the whalers had set off across the Arctic in search of bigger profits. Later, Norwegians arrived and began hunting seals, Arctic fox, and polar bears. By the late nineteenth century, Norwegians were killing more than 400 bears a year, many at unmanned sites where any animal that tugged on a baited gun would trigger a blast that killed or more often maimed it (a technology invented in Svalbard and copied around the world).

The last phase of Svalbard's history now arrives, the one that confronted me the moment I touched down at the airport: the coming of the mines. A little over a hundred years ago, a rush for coal and minerals began, with hundreds of newcomers of all nationalities trying to establish claims. With Svalbard an international no-man's-land, that often meant little more than putting up a sign asserting ownership: the museum has a colorful collection of them. English, Norwegian, and American entrepreneurs all opened large mines in the 1900s. The American John Longyear established Longyearbyen alongside the Norwegians. The Russians later bought Barentsburg, where they remain, from the Swedes.

For just over 300 years the islands had been exploited, fought over, and occupied by citizens of a dozen nationalities, but never ruled by anybody. Nowhere else in the world was quite like this no-man's-land. Finally, in the aftermath of the First World War when so many maps were redrawn, the Svalbard Treaty was signed in Paris in 1920,[10] giving Norway sovereignty over Svalbard. But it is a very odd form of "absolute and unrestricted sovereignty" that continues to connect Svalbard to its unruly past.

While Norway is in control, all other signatories to the treaty acquire rights to fish, hunt, trap, set up mines and commercial enterprises, and acquire mineral rights in Svalbard. Anyone from the now very long list of treaty signatories—including nations such as Saudi Arabia, Afghanistan, and Venezuela with no particular links to Svalbard—can go and live there and start work. Norway has the right to regulate their activities but not to profit from them. Any taxes collected in Svalbard must be spent there, which makes taxes lower than in Norway proper. No one is allowed to use

Svalbard for any military purpose and, finally, under a special treaty obligation, Norway must preserve the natural environment.

One other peculiarity of the treaty matters to some. As I was told firmly on a visit to one of the Russian mining towns, "The Soviet Union was not invited to the Paris conference." In 1920 the Europeans still hoped the Bolsheviks might be defeated. Although the Soviet Union signed the treaty in 1934, that was part of a larger set of diplomatic moves that gained it recognition from the United States. The Russians still don't believe the treaty was truly fair to their long connection to Svalbard and its proximity to their shores. Svalbard's current population of 2,400 is 55 percent Norwegian and 45 percent Russian and Ukrainian, with the Russian government committed to increasing its activity on the islands.

Still, it is thanks to this odd form of sovereignty that I was able to see scientists of every nationality working happily together at Ny Alesund and to pass Russian mining communities around the islands. Thanks to the treaty, there are nature reserves everywhere and no one would dare think of shooting a bear. And there is no sign of the military, although the islands were briefly occupied by Nazi Germany in World War II.

So has Svalbard's history of competition between nations and the unsustainable exploitation of wildlife been transformed into its opposite: a model of cooperation between nations and environmental protection? That is what my scientist friends at Ny Alesund like to think. But as I stayed a little longer, I found tensions below the surface.

A chat with long-term local residents in a Longyearbyen bar is revealing. Normally I wouldn't enter a bar in Norway because of the risk of bankruptcy, but the Paris Treaty provides tax-free shopping. Conversations with a representative 0.5 percent of the town's population (seven people) brought up the same points again and again. Illegal fishing by Russian boats always comes first.

"The Russians are much poorer than us. That's why we are constantly arresting their fishing boats for illegally fishing here. They have nowhere else to go to make money quickly, and anyway they don't believe it's illegal," says one part-time fisherman.

A man who works as a mountain guide complains: "If the Russians break the rules, not much happens even if they are arrested. The fines don't get paid. As soon as any problems go high up, politicians don't want to hear about it. Norway is really run by StatoilHydro [the giant oil company], and they need to get a share of Russia's Shtokman field. Everything will be sacrificed to plans for Arctic oil. Norway is a petrostate just like Saudi Arabia."

His companion adds: "We Norwegians pretend to have real sovereignty here but we know we don't. Instead we use the treaty to blanket the islands with nature reserves. That is the only way we can keep people out and stop the Russians building more mines and roads. It's a trick."

A quick trawl through the local newspaper's database puts some substance behind these opinions. Sixty percent of Svalbard is covered by nature reserves—and they are patrolled more strictly than in Norway proper. Arrests of Russian trawlers operating illegally in Svalbard's waters are quite frequent. In one case in 2005, two trawlers were stopped and brought into Longyearbyen, while another just sailed off with two Norwegian inspectors on board. They were later returned by the Russian Navy, but charges of "kidnapping" were simply abandoned three years later.

When I visited in August 2008, five Russian ships had already been detained that year. Earlier, Russia's own fisheries inspection vessel, the *Mikula*, put in at the Russian mining town of Barentsburg. The Norwegian coast guard and police promptly showed up and told the ship to get out of Norway's waters. The Russian captain accused the Norwegians of violating the Treaty of Paris. He had a point. That 1920 treaty states that signatories "shall enjoy equally the rights of fishing and hunting." So what is going on?

Norway's position is that the treaty applies only to the islands and their immediate waters: the wider 200-nautical mile exclusive economic zone (laid out in the Law of the Sea) belongs to Norway because it is an extension of Norway's own continental shelf. The fisheries surrounding the islands and rights to the sea bottom thus belong to Norway. Russia does not recognize this claim—and other treaty signatories do not exactly agree either.

In a nutshell, the argument is that when Norway was given sovereignty back in 1920, Svalbard's territorial waters extended only three miles from the shore. If Norway wants to extend Svalbard's territorial waters under the much more recent Law of the Sea, then those waters should fall under the terms of the treaty, and all signatories enjoy equal rights to what is there. Norway wants Svalbard itself to be dealt with under the treaty but wants the surrounding waters to belong exclusively to Norway. Of course, none of the diplomats will put it all quite so plainly (no one dares make an on-the-record statement) as no one wants to start a quarrel over an issue that might be better left alone.

The search for oil and gas in the Arctic threatens to raise the stakes. In 2008, Russia unexpectedly announced that it was planning to send survey ships to start exploring Svalbard waters for hydrocarbons. Jonas Støre, the Norwegian foreign minister, replied with a statement that neither Russia nor any other signatory to the treaty had any right to carry out industrial activities in the area. He repeated the argument that this offshore area is part of Norway's continental shelf and the Svalbard Treaty is not relevant. Meanwhile Norway's own oil exploration companies are demanding access to the Svalbard waters. Nations have argued over Svalbard's whales, bears, walrus, and coal—is oil next? Those survey ships have not appeared yet, but when the price of oil goes up again, no one can be sure. The diplomatic situation is delicate and carefully balanced. Norway is very anxious to keep good relations with its large and ambitious neighbor. As the denizens of the Longyearbyen bar understand, StatoilHydro has taken a 25 percent stake in the ambitious Russian Shtokman gas field project not far away and sees success as key to getting a cut of further riches in Russia's Arctic waters. Not far to the east of Svalbard, there is that still troublesome and unsettled maritime border with Russia that makes Norway cautious. Of course, Norway has many cooperative initiatives with Russia in their shared Barents Sea, and you can certainly say that relations are at their best since the Cold War.

A few months later, back in Norway, I heard Rear Admiral Trond Grytting, chief of the regional military crisis headquarters in Bodo, located on the north coast of Norway across from Svalbard, lecture on the High North military situation. He recalled the Cold War days when there were NATO bases in northern Norway, Greenland, and Iceland which faced Soviet bases in the Kola Peninsula. Both sides were ready for an invasion by the other. "This is now history," he said. "However, it would be naive to rule out the possibility of incidents involving the use of arms and show of power in our region. It is pretty much a fact that enormous amounts of biological and geological resources in combination with unsettled borders are the most likely source for a conflict in any region."

I am afraid my scientists at Ny Alesund are a little naive to believe that international cooperation is a certainty. This is a difficult region, and old rivalries and competition for resources don't go away so easily. Tensions elsewhere in the world can echo up here too. When Russia sent troops into Georgia in 2008, worries about conflict in the High North rose rapidly. Norway has to be careful not to annoy its neighbor and other nations that are usually its friends, even though they don't share Norway's interpretation of the Paris Treaty. But I don't feel pessimistic. Leaving Longyearbyen, I had a last drink in that bar. This time there were lots of Russian tourists there, too, and everyone had their glasses raised to a toast of "Russia, Norway—Eternal Friends." If the alcohol stays cheap enough, Svalbard should do well.

ANIMALS

Chapter Nine

TROUBLE AT THE TOP

Out on the sea ice, polar bears rule and they know it.[1] In August 2008, near the islands of King Karl Land in Svalbard, I watched a female bear walk across the sea ice to our 400-foot-long ship and try to climb straight on board. The side of the ship was a little too high for her, so after half an hour of nibbling the ship's bow and playfully scratching its sides, she tried a different strategy. She lay down on the edge of a nearby ice floe, gave a long yawn, folded her paws under her chin and fell asleep. That is to say, she closed her eyes and looked as though she were asleep, but there was something slightly suspicious about her cocked ears.

Patient "still hunting" at the edge of a floe is the polar bear's number-one technique for catching seals. A bear may sit or stand like this for an hour or more, utterly still but alert, until the instant a seal surfaces.[2] Then comes a flurry of lethal action. Knowing this, I was not much inclined to see if I could climb down onto the ice to take a close-up photograph, beautiful though she was. This was a bear that was truly in her element; hunting a large ship suggests a swagger you won't see in any other large mammal.

Eventually she grew bored, stood up, and strolled off into the far distance, across a vast patchwork of broken ice floes, some not much bigger than herself. Where there was a small gap, she would spring onto the next floe; where it was larger, she would slide into the water, paddle for a while, then smoothly slither out onto the next floe. Watching her progress, I was reminded of just how strange a creature she is. A polar bear really is a great big land animal that has figured out how to walk on water and travel far out to sea. If the temperature rises just a few degrees and the ice vanishes, then the bear will be helpless. Bears can swim a very long way, perhaps fifty or even a hundred miles if they are truly desperate, but they are not whales.

In the area where I saw that female, polar bears were shot for fun by tourists from passing ships not long ago. But since the 1970s, the Norwegian government has been protecting polar bears with such seriousness that the locals joke that "you are better off shooting a man than a bear: the authorities will investigate you less thoroughly." These days, only scientists are permitted to go ashore on King Karl Land, so as not to disturb the female bears who come here to build the dens in the snow where they will give birth.

Everywhere in the Arctic the polar bear is protected now. In Alaska, where "sport hunters" used to shoot polar bears from airplanes, the polar bear was listed as a threatened species in May 2008. Russia long ago banned polar bear hunting in its part of the Arctic, but recently added a quota for indigenous peoples to help control poaching. Canada and Greenland also protect bears while granting a small number of hunting licenses to indigenous people. All the circumpolar nations work together under the terms of the International Agreement on the Conservation of Polar Bears, a remarkable agreement signed in 1973 when the Cold War was still on. Other things being equal, polar bears should be making a continuing comeback across the Arctic as hunting pressure recedes.

There are between 20,000 and 25,000 polar bears living in the Arctic now, and in some areas their numbers may well be growing. Sadly there is another pressing danger, which will wipe out any small gains the bears are making. That is, of course, the rapid disappearance of the sea ice. Pictures of an unhappy bear perched on a tiny, melting ice floe floating on a blue sea have become the symbol of an Arctic under threat. The pictures highlight the strange irony that the region's top predator, a terrifying and powerful hunter, is helpless without the ice. Other iconic animals—including the walrus, the white beluga whale, the narwhal with its extraordinary tusk, and the mighty bowhead whale—also may be at risk, as they all habitually live on, near, or under the ice. Then there are the seals of the Arctic, especially the ringed seal which the bear relies on heavily for its food, and the bearded, spotted, ribbon, harp, and hooded seals. All too have a life connected to the ice.

Polar bears look as if they are land animals that have only recently taken to the sea, because that is exactly what they are. Polar bears are descended from brown bears trapped in the far north when most of North America was covered in ice during the last glacial period. They could no longer find their usual berries, fish, and occasional caribou to eat, and turned to seal, perhaps first scavenging dead seals that were washed up onshore and later hunting newborn pups on the accessible ice close by.

Over time, the bear changed. Compared with its relatives to the south, a polar bear is bigger. At 1,300 pounds for a big adult male, only the giant Kodiak bears compete for the title of the world's largest land carnivore. And its coat has, of course, turned white. The polar bear's best-kept secret, though, is that underneath the white fur its skin is black, a secret occasionally revealed in the black scars on animals that have been in a few fights. Having a black skin under a white coat is a clever trick. The white fur reflects back all the light that we and other animals can see so that the bear is well disguised against the ice. Longer-wavelength light, of the kind we can't see, shines straight through the bear's fur and is soaked up by its black skin, helping keep it warm. Despite these and other changes, including broad paws to help it swim, bigger and better-insulated feet, and a stronger, more mobile neck, polar bears and grizzly bears remain close relatives. Occasionally they mate in the wild, and grizzly–polar bear hybrids may even be growing more common as polar bears are forced to spend more time on land.

Even with these adaptations, polar bears could not live on the frozen ice, where temperatures may fall to -50°C, without eating the ringed seal's rich blubber. Ringed seals in turn eat fish from the rich and warm (relatively speaking) waters beneath the ice. The ringed seal is a beautiful creature too. The back of its coat shades from light to dark gray, and it is dappled with circles of silver surrounding a darker center that give it its name. It weighs about 150 pounds fully grown, which is a couple of weeks' worth of blubber-rich food for a bear. Seals, of course, don't want to end up as a bear's lunch. Ever since brown bears began moving onto the ice, bear and seal have been running an evolutionary race, one refining its skills at finding seals to eat and the other at avoiding bears.

On a visit to Greenland, I put myself in the bear's place, going out to hunt ringed seal with Dines Mikaelsen, an Inuit hunter who lives in Tasiilaq in eastern Greenland. Mikaelsen is a young man with a bright smile and a ready laugh; he can speak several languages and lives in the center of Tasiilaq, his home surrounded by his dog team. I would not step among these fierce Greenlandic huskies if I were not with him; he can call each one separately by name and tell them what to do, and each obeys. As the top dog, he is careful not to provoke jealousy, and if he gives one dog a hug he goes around and hugs every one. He is proud of his hunting skills and has a big photo of himself with his father when, as a teenager, he shot his first polar bear.

Seals are taken both for food and for their skins, which are made into clothing at a workshop in town. Hunting a ringed seal is not so easy. It was July when we set out amid the floes and occasional icebergs in a small, open white boat. Now and again Mikaelsen would spot a distant seal which would poke its head above the water and quickly disappear; usually it was gone while I was still searching for it with my inexpert gaze. Many cold hours passed before we saw a single seal on a floe in the distance.

Mikaelsen cut the outboard motor after giving it a quick last rev to push the boat toward the seal. We crouched low and kept silent as the white boat drifted slowly toward our quarry. I could see that the seal never rested for long. Even here, in the middle of the ice, it would still raise its head every few minutes and briefly scan the horizon for predators.

Many tens of thousands of years of being hunted by bears have given the ringed seal a conspicuous wariness. In Antarctica, where seals face no predators on top of the ice, I have come right up to an indifferent sleeping Weddell seal. Here in Greenland though, stealth is vital. Our seal had begun looking at the white boat, perhaps wondering if there was something odd about this piece of drifting ice, when Mikaelsen shot it cleanly through the head with a rifle, and it slumped down dead.

We took the seal home, Mikaelsen quietly giving it a pat and saying a word of thanks. The seal replied with a last reflex, urinating into the boat.

A polar bear would have to sneak up much closer than we were to have any chance of catching a seal. That is why "still hunting" next to a seal's breathing hole in the ice is the best plan. As the summer seas freeze in winter and there are fewer "leads" (patches of open water among the ice) in which to surface, ringed seals use the heavy claws on their front flippers to keep their breathing holes open. This adaptation, which they are the only Arctic mammal to possess, enables them to live even in seas that have totally frozen over.

Each seal usually hedges its bets by keeping open three or four holes, more than 600 feet apart. In early winter, the holes will be hidden beneath a crust of fallen snow. As the snow deepens, seals will build lairs in the snow above their breathing holes so that they can rest and in the spring give birth to their pups.[3]

A polar bear can sniff out these secret places from a long distance, perhaps a little over half a mile, and will sit or stand alongside one, waiting with total concentration.[4] When the seal comes up to breathe at one of its holes, it is sensitive to the slightest tremor in the ice above and will dive immediately if it detects danger. This is the crucial moment. If the seal surfaces at the wrong hole, the bear will strike with great speed, smashing through the snow's crust, grabbing the seal by the neck, and hauling it out onto the ice.[5] It is all over in a second.

The bear quickly strips the seal's skin away to reveal the great store of blubber. News of the kill instantly spreads: out on the ice, you can track down a still bloody bear, standing over its prey, by the crowd of noisy circling ivory and glaucous gulls that suddenly gather from nowhere to snatch bits of seal.

The polar bear does not find it any easier to hunt ringed seals than we did. Over a year, there are a few times of feasting and many long periods of fasting. In summer especially, when the sea ice has retreated and there are many open leads where seals can surface safely, bears can have a tough time. A bear may kill forty to fifty ringed seals a year. Half of those kills are pups still in their birth lair in spring. Another third of the kills are naive young pups taken in early summer. The rest of the year, the pickings are very thin, and polar bears must depend on the fat that they have built up during the good times.

The ability of the polar bear to withstand long fasts exceeds that of any other mammal. During that six- to eight-week period in spring when food is abundant, polar bears can triple their body mass. At its end, 50 percent of their weight will be fat; after the lean season that will fall to just 10 percent.[6]

Mother bears with new cubs are especially dependent on that spring feast of ringed seal pups. Once a female becomes pregnant, she will eat as much as she can to pile on fat and then retreat into a den in October and wait out the winter. There she will give birth, usually to twins in December or January, and nurse them there through to late February or even mid-April. By then, each cub will weigh around twenty to thirty pounds, while their mother will be hundreds of pounds lighter after having eaten nothing for four to five months. When she finally emerges she is ravenously hungry and must quickly lead her cubs onto the shore ice where ringed seals pups can be found.

The lives of the polar bear and the ringed seal are thus entwined with one another, and both also depend utterly upon the ice. For the bear, the ice is a platform that allows it to walk on the sea and hunt the seals. For the seal, the ice is a place to rest, hide, and raise young while giving access to the rich waters below.

With the end of the summer ice and open water throughout the Arctic now a real possibility, scientists are trying hard to predict what will happen to the bears, given their dependence on ice. We can start to see answers from the bears that live in Canada's West Hudson Bay. Here, south of the Arctic circle, the sea ice always melts away in summer and the bears have to spend the middle of the year on land, hungry, waiting until the bay freezes over again and they can get back to hunting seals. The bears gather around the town of Churchill, "the polar bear capital of the world," which has a tourist industry built around trips to see the bears aboard giant "tundra buggies."

Steven Amstrup of the U.S. Geological Survey (USGS) in Alaska has spent thirty years studying polar bears around the Arctic, including those in Hudson Bay. "It makes it easier to study the bears, as they are a sort of captive audience onshore rather than being spread out over thousands of square miles of sea ice," he explained. "There is an easier change in the environment to look at. Here, the ice is simply melting earlier. There are no worries about how much of the remaining ice is multiyear ice, about the quality of the ice, and all those other things that make it much more complex to study bears in the polar basin."

Through patient study of bears and cubs of all ages, Amstrup and his research student Eric Regehr have shown that for every week earlier that the sea ice breaks up, survival rates decline between 3 and 8 percent, depending on the sex and age of the bear.[7] "They are coming ashore a lot earlier, and they are coming ashore at a lighter weight," says Amstrup. "That means they produce lighter cubs, and lighter cubs apparently don't survive as well."

Elsewhere in the Arctic, where the bears don't spend summer onshore, the research is more difficult, but the bears seem equally threatened. In the Beaufort Sea, for example, Amstrup has found that fewer cubs are surviving their first year of life. "The decline fluctuates with the sea-ice condition. It follows years when ice was away from the shore for an extended period," he says. "We don't know why, but it might be that there is a lot more open water up there, and we know that cubs don't do very well swimming, so they might be more vulnerable to drowning or hypothermia. In recent years, the ice has been much rougher than it used to be, with sharp, angular shards of ice sticking up in all directions. That might make it more difficult for those tiny cubs as they try to follow their mother who needs to feed."

In 2008, the USGS pulled together computer models of the future state of the ice, the kind of models that we have seen already, with what was known about the life of bears. The goal was to predict the future of polar bears across the whole of the Arctic.[8] There was no good news; only the grim conclusion that two-thirds of the world's current polar bear population would be lost by 2050. But that was enough for immediate political action. In the United States the polar bear was quickly given extra protection as a "threatened" species.

Reading the USGS's gloomy forecasts, I could not help wondering what would happen to that confident female bear that I had seen in Svalbard, strolling across the ice to hunt a ship. Her part of the Arctic fell into the worst category in the USGS report: "extirpation by 2050." That means her grandcubs may be the last polar bears to live there.

Bears will be worst hit where the sea ice shrinks so far away from the shore that they cannot reach the ice to hunt without an exhausting swim, or cannot make it back to the shore when it is time to dig a birthing den in the snow. Polar bears are powerful swimmers, but the water is not their real home and the farther the ice retreats from the shore, the greater the risk that polar bears will drown. In a single day in August 2008, scientists in a plane spotted nine bears swimming far out in the Bering Sea off the coast of Alaska. The ice edge had melted away far to the north, and one of the bears was already more than sixty miles from the shore. As the seas widen, not only do the distances that bears have to swim increase but so do the size of the waves, increasing the risk of exhaustion and drowning.

Polar bears will have a better chance in parts of the Arctic where summer ice may hang on longer. Along the coast of the Canadian islands and around to Greenland, where the ice circling in the Beaufort Gyre tends to pile up against the shores, the USGS prognosis for bears gives them an extra thirty years, with "extirpation by 2080" expected. Only in one region, in the narrow, frozen channels among the northerly Canadian islands, might the bears survive until the end of the century.

These are grim forecasts, but the USGS rightly says they are "conservative." They are based on those computer models that, as we have already seen, do not succeed in keeping up with the true speed of the collapse of the ice. If the more pessimistic models, like that of Wieslaw Maslowski, which predicts ice-free Arctic summers by 2013, turn out to be correct, then polar bears are going to be in very serious trouble within the next ten years. Just like those forecasts about the sea-ice area, we may have to move forward the forecasts about bears by fifty to sixty years.

The ringed seal is not nearly as photogenic as the bear, and it is harder to excite the public about its fate or to find research grants to study it. Even so the U.S. Center for Biological Diversity has petitioned for a "threat-ened" or "endangered" listing under the U.S. Endangered Species Act for the ringed seal, along with the bearded, ribbon, and spotted seals, and the walrus. For the seals, an official review is now under way. In early 2009, small unmanned aircraft were used for the first time to survey the ice off the coast of Alaska and begin a seal population count.

Shaye Wolf, a marine biologist at the Center for Biological Diversity, is leading the charge. She has already seen the power of science when her own seabird studies in Baja California helped to stop the development of a huge liquid natural gas terminal. Now, she believes that early warning sig-nals for the ringed seals and other marine mammals are coming in loud and clear.[9] "The same studies that show that polar bears are suffering as the Hudson Bay ice breaks up earlier each year also show that ringed seals are affected," she explains. "The sea ice breaks up before the pups are fully weaned, so they have to leave their mother before they are fully developed and enter the water."

Across the Arctic in Svalbard, Wolf says ringed seals are failing to breed as the ice disappears northward. To find out more, I tracked down Kit Kovacs, a Canadian who studies whales, seals, and bears at the Norwe-gian Polar Institute in Tromsø.[10] "We haven't seen ice on the west coast of Svalbard since 2005," she said. "For as long as we've been looking, the west coast of Svalbard has been one of the best areas for ringed seal reproduc-tion. Now there has been virtually nothing since 2005." Nothing at all, I said in astonishment. "Correct," she replied.

Maybe the seals have moved farther north where there is still ice? But that turns out to be a forlorn hope on my part. "Up in the northeast cor-ner of Svalbard, there has been ice," she says, "but it's been very unstable, moving ice, not the kind of ice that ringed seals need. They need land-fast ice or large free-floating pack ice that isn't going to crush them. They need a stable platform for a minimum of eight weeks to get the business of reproduction done. So in 2010, the baby seals that should be coming into the population as adults, breeding for the first time, simply won't be there."

Ringed seals live all around the Arctic and there are still many mil-lions of them.[11] But these observations from Svalbard show that the moment they lose the ice that they need to give birth to their young, their future becomes bleak.

The walrus—which comes in separate Pacific, Atlantic, and Laptev Sea varieties with the Pacific walrus making up 90 percent of the worldwide population of around 300,000 animals—is the Arctic's most charismatic big beast after the bear.[12] Its great bulk (Pacific males may weigh close to 3,000 pounds), thick skin, and fearsome tusks make it secure even from polar bears. An injured animal or a pup might be attacked, but an encounter with an adult walrus will usually go badly for the bear. That might be why a big walrus seems so self-confident. A sleeping walrus on an ice floe barely bothers to raise its head for a few passing humans and just slumbers on like a giant slug. A solitary human close to a walrus beach may excite just a little curiosity. I've sat and waited on a beach while a walrus or two hauled themselves up closer and closer to take a better look at me with their watery, lugubrious eyes. (Curiously, they seem to find me more interesting if I sing loudly. The same trick works well for seals too.) I know it is time to move off when they are so close I begin to smell their notoriously bad breath. The fringe of sensitive whiskers around their mouth enables them to find shellfish by touch in the dark and muddy sea bottom, but their diet comes with a distinctive smell; I've only encountered worse when paddling downwind of a sleeping humpback whale.

Walrus may be large, strong, well-armed, and smelly, but they too are dependent on ice. In summer, Pacific walrus females and their pups leave the males and travel with the ice edge as it retreats north, through the Bering Strait and into the Chukchi Sea. The ice is a perfect moving platform for mother and young cubs, who nurse for at least a full year. As the ice edge heads north, it carries them across the shallow seas where there are plenty of clams to eat.

Or at least, that is what used to happen. Now, in bad years, the ice is retreating so far north that it ends up in the deeper Arctic Ocean where there is no food to be found. At the USGS's Alaska Science Center, Chad Jay has been following walrus movements. In part, he explained, that means creeping up on a walrus armed with a crossbow and firing a dart equipped with a radio transmitter into its blubbery back. That does not worry the walrus too much: the skin on an adult is about an inch and a half thick and there are another couple of inches of blubber underneath that. And as Jay can hit a walrus from thirty feet, it is not too scary for the animal or for him.

Jay has been seeing that when the ice retreats far in the deep Arctic, walruses are forced to abandon the ice and try to make a living from the shore. The result can be catastrophic. "We saw upwards of several thousand female walruses and their young coming to shore in Alaska in 2007. The same thing happened in the north coast of Chukotka on a much more dramatic scale where tens of thousands of walruses took to land. Russian colleagues saw stampedes into the water and often the little ones got trampled and died," he told me.

If the ice vanishes even earlier, during the pupping season, an adult walrus may simply have to abandon its pups. That seems to have happened in 2006 when researchers came across the sad sight of a group of crying baby walruses far out to sea off the Alaskan coast. All were too young to survive on their own and probably drowned. That year the sea had been especially warm. Normally, a walrus mother keeps her pups close by, leaving them on the ice only when she has to dive for food to the shallow seas, or when the pups have to rest. If the ice melts away too rapidly, the pups may not be able to keep up with its retreat, or the ice may be too far from places the mother can go to find food. Sometimes, on a long swim, a mother may carry her pup on her back, but there are limits. The pups may have to be left behind.

Carin Ashjian, a marine biologist at Woods Hole Oceanographic Institution who studies whales and the plankton that they eat, was one of the team of researchers on board the coast guard cutter *Healy* that came across the pups. "We are not walrus scientists. It was an accidental discovery that we wanted to get out as soon as we could. As far as we know this is something totally new. No one has ever reported seeing abandoned baby walruses out in deep water before," she told me. With just one observation it is hard to be sure if it is a sign of what is coming, but the team concluded that if the walrus "cannot adapt to caring for their young in shallow waters without sea ice available as a resting platform between dives to the seafloor, a significant population decline of this species could occur."[13]

We are even less sure of the risks to the Arctic's three whales: the bow-head, the narwhal, and the beluga. "It may sound strange," says Kovacs, "but although the whales are all closely associated with ice, we can't be sure if ice is essential to them or not."

Traveling off the coast of northern Greenland one August, I had the good luck to see narwhal surface among the ice. For a brief moment, three improbably long spiral horns broke through the water and waved above the sea like magic wands. One animal twisted around and, for a second, his gray, wet body glistened in the low sunlight. Then they all dived and were gone. I stayed for an hour searching the sea for them, but they never returned and I have not seen one since.

Strange though it may seem, we still do not know what that long spiral tusk is for, so many centuries after it was first passed off as the horn of a unicorn. As early as 1250, the Norwegian book *The King's Mirror* recorded how it was possible to grow rich trading their tusks from Green-land.[14] The royal scepter in England is made from narwhal and so is the frame of the Danish royal throne. But what the narwhal needs that tusk for is less clear. Biologists have proposed that it could be a tool for digging the seafloor or breaking ice, a weapon for fighting other narwhal, an instrument of hearing, or even for spearing fish. I tried asking Kristin Laidre from the University of Washington in Seattle what she thought. She spends her summers in Qaanaaq, in the very far north of Greenland, tagging narwhal to try to find out where they go and what they eat. "It's clearly not used for anything critical to survival," she said, "because only males have a tusk." She has observed narwhal at the surface, crossing their tusks "in a gentle, almost balletlike manner, usually there are two or three males with a female very close to them." She sees the tusk as analogous to the peacock's showy tail.

The bowhead is well adapted to the Arctic, with a huge, strong head that allows it to break through ice and thicker blubber than any other whale. It also has more baleen—the hairy plates around the whale's mouth that it uses to sieve food from the water—than other whales as it relies on mopping up patches of plankton from the Arctic seas. Its population seems to be slowly growing now as it recovers from centuries of human hunting, but its numbers are still small with only 8,000 to 9,000 left in the whole Arctic.[15]

Beluga and narwhal both like ice, too, but it can also be a terrible enemy to them. Schools of these whales are sometimes trapped by ice in small areas of open water, called *savssats* by Inuit hunters, and cannot swim far enough to reach the next hole where they can breathe. As the freeze continues and ice closes in on them, indigenous hunters and polar bears take advantage of this unexpected feast.

All three of these whales may like to be near ice only because it helps protect them from predatory killer whales. Amid dense pack ice, killer whales are at a disadvantage because their enormous sail-like pectoral fin makes it harder for them to find a place to surface.

"Beluga can probably use ice as a refuge to hide out," explains Rod Hobbs, a biologist from the National Marine Mammal Laboratory in Seattle who has been tracking beluga around the Arctic, especially the small population living in Cook Inlet near Anchorage.[16] "Beluga are one of the few cetaceans that have a fairly well-articulated neck," he said. "They can tip their head forward into a position more like a human head with their blowhole at the top. That means that they can get into fairly narrow spots to breathe." Beluga have been tracked off the coast of Alaska as far as 83° north latitude, where the ice cover is 90 percent, although that is not so common.

Narwhal really seem to love lots of ice, but that might be because they are very good at finding food under it. "We don't know if they go out to the center of Baffin Bay in winter because they like to be in 99 percent ice cover, or because it is a good place to feed," says Kristin Laidre. "My guess is that it is probably a good place to feed and that narwhal essentially evolved to exploit a habitat that no other marine mammal can exploit, and so they have it all to themselves. There are predictable concentrations of prey so it's a very smart strategy. If you can adapt to that environment it is a good place to go."

As the Arctic changes, Laidre thinks that the narwhal will find the future tough. "It has a relatively small population worldwide, and it is a real specialist in terms of what it eats and where it goes: this doesn't really

seem to suggest that the animal will be able to adapt to change." Beluga
are similar in size to narwhal, but they are more easily able to live away
from ice. "They can swim up into the dense pack ice like narwhals, but
they also live in muddy estuaries. They can swim up rivers, live in ice-free
areas and feed on anything. They will eat whatever swims by, whereas we
have found only four or five prey species in the stomachs of narwhal," she
explains.

Hobbs agrees. "They can live in areas of a lot of ice, but in Cook Inlet
there's ice in the winter but not in the summer, so they are adaptable to
areas without ice. They seem to do well in rivers where salmon and eula-
chon run. As the Arctic warms, some rivers with salmon runs will decline
but others farther north may pick up. I think beluga will be able to find a
place to go."

I'm not really too surprised. There are thought to be around 120,000
beluga spread over the Arctic but only 55,000 narwhal,[17] most of them living in Baffin Bay: as change comes and all its special creatures face the
cruel mantra of "move, adapt, or die," beluga may turn out to be the lucky
generalists.

In 2008, a team of six well-known Arctic marine mammal researchers tried to put together everything they knew about the different factors
threatening the big mammals in the region, with the vanishing ice as the
biggest stress, and to work out which creatures were most at risk.[18] Just as
we might expect, narwhal and polar bear come out as most sensitive to
change, then bowhead, beluga, and walrus.

Laidre was one of that team along with Steven Ferguson, an evolutionary ecologist from the University of Manitoba who has spent the last
couple of decades studying many of the big mammals of the Arctic. Like
many Canadian researchers now, he works closely with Inuit hunters who
have often seen more of the animals than the scientists who come to study
them. I asked Ferguson if there were any gleams of hope, especially for
the ringed seal. I was to be disappointed. "I think all of the ice-dependent
species are in trouble, in quite big trouble if we lose ice. But if we have to
compare among them then I think narwhal and polar bears are probably
ones that we should be most worried about. But definitely, ringed seals
and bearded seals are really going to be in trouble as we lose the sea ice.

We tried to compare the different ice-adapted species but generally the future doesn't look good for any of them."

That sad conclusion is not the end of the story. I can best explain by turning outside of science to a pack of medieval tarot cards. It's a pack I know well, having earned a little money as a fortune teller when I was a student (a safer bet than playing poker). In that pack there is one card that you would prefer not to turn over during a reading. That card is "Death." It shows a grinning skeleton with a scythe in a field littered with limbs and a severed head with a gold crown. How that card was interpreted in the fifteenth century I hate to think, but nowadays it is no longer read as an "end," but as a transformation, a new beginning. There it joins in spirit another card, "The Wheel," which shows beasts on a great wheel, some falling while others rise.

The Arctic is undergoing a transformation on a great scale, larger than anything that has been seen on this planet in human history. The king of the Arctic, the polar bear, is going to lose his crown, and be mourned, but that will not be the end. Another creature is coming to replace him at the top of the food chain in the new Arctic.

Chapter Ten

THE BOTTOM OF THE WEB

Every ship I have traveled on through the Arctic ice has always had a few keen naturalists on board. They'll be up on deck, night and day, scanning the horizon with their binoculars, and quite a few will have a camera with a giant telephoto lens at the ready. Usually I am up there with them as there is a special thrill to spotting a distant bear or a walrus lying on the ice like a big, fat contented slug. Passing a walrus you'll always hear the same joke about its penis bone (the baculum), which gives it a permanent erection—a fact that the male walrus finds hard to conceal. Then there are the birds that come along unexpectedly: murre flying home in a straggling line, a group of tiny dovekie settling on the ice, or a long-tailed jaeger performing acrobatics overhead. (Or was it a pomarine jaeger? There will always be someone to quickly correct you if you make a mistake.) The distant fog of a whale blow gets people most excited of all: did the shape of the plume and the brief flash of a fluke mean that it was minke, fin, humpback, or something really rare? We all search the distant seas hoping to see it again and be the first to shout, "Blow!"

Sometimes I try to persuade my shipmates that there is something else to see besides the birds and the beasts. You can get an occasional glimpse of another side of Arctic life if you look down at the floating slabs of ice as they are pushed aside by the ship's bow. The surface of the ice will be pure white, shading into blues and grays depending on the light, but as the ship passes by and the ice floes rock back and forth, you'll sometimes catch a flash of pale browns and deep greens hidden under the water line. These are algae growing on the ice. If you can persuade someone to haul a block of ice on board and crack it open, you'll see that the stain of color runs up inside the ice as well.

We are seeing the source of all life in the Arctic.[1] The algae growing inside and under the ice, along with many more that live in the sea where the ice has turned into open water, capture the sun's energy and power everything that lives among the Arctic seas.

This is the bottom of the Arctic food chain, the base of the pyramid on which everything else rests. Zooplankton eat the algae, fish eat them in their turn, and so on upward to seals and birds and polar bears and to man, too, for two areas of the Arctic, the Bering Sea and the Barents Sea, are among the world's most productive fishing grounds. Any change that happens down here, at the bottom of this web, may alter the shape of the whole web of life in the Arctic seas.

That is as much as I can explain to my naturalist shipmates, who say "how interesting" and turn back to the horizon to look for something bigger and more important, leaving me with a block of stained ice which has been judged unfit for use in the ship's bar. I'd need some equipment, a microscope at minimum, to take them onto the next step. If I could zoom in on the inside of that block of ice, I could show that it is riddled with a labyrinth of tiny tunnels. Inside the ice there is an unseen and wonderful world that they would find hard to believe. Within it are algae, tiny grazers that eat the algae, and the tiny predators that eat the grazers in their turn. It is a microcosm of the larger world above that my shipmates are looking at with binoculars.

The profusion of life inside the ice and on its undersurface is one of the Arctic's best kept secrets. A lump of sea ice may look much the same as ice that you take out of your refrigerator, but it is not. Sea ice began as salty water. As it freezes, salt is squeezed out of the growing ice crystals, and left surrounding them as a briny liquid which is just too salty to turn into ice. Instead, the brine flows into a network of tiny channels which gradually interconnect and broaden as they flow down to the bottom of the ice and reach into the sea beneath.

A few years ago I heard about a remarkable experiment carried out at the Alfred Wegener Institute, Germany's main polar research institute. Scientists took a lump of sea ice and spun it in a refrigerated centrifuge (the ice scientist's equivalent of a spin drier) until all the brine drained from it. Then they filled the ice with a special resin which they could harden by shining ultraviolet light on it. After the ice was allowed to melt away, they were left with a beautiful resin cast of the branching channels, big and small, within the ice.[2] Powerful microscopes showed that the insides of the ice were as complex as the interior of a sponge: there were broad routes, several millimeters across, leading down to the bottom of the ice, and vast numbers of narrow, interlinked byways no wider than the finest fibers of a spider's web.

Rolf Gradinger, one of the team who made that cast back in 1991, is now at the University of Alaska in Fairbanks. When I called him up to ask him about his work, I was thrilled to find that he is still as madly excited about sea-ice tunnels as if he had discovered them yesterday. "That three-dimensional cast was really new. It was so amazing to see how connected everything was," he says. "It was just an eye-opener to see that for the first time. Whenever I work on sea ice, I have that picture in my mind."

Many of the tunnels in the ice are lined with algae, often along with dense mats of bacteria. The algae themselves are remarkable. They live in a dim, twilight world. "They are unique microbes on the earth," says C.J. Mundy, an ice algae researcher at the University of Quebec at Rimouski. "The algae are so well acclimated to the faint amount of light that they are able to increase the amount of chlorophyll, their photopigment, to collect more of the light." Down in these tunnels where the algae grow, strange-looking grazers with exotic names come to feast.[3] There are rotifers, with a bloblike body driven by a crown of lashing cilia and flagellates which, as their name suggests, are powered by whiplike flagella. These in turn end up as prey for tiny carnivores, including the wormlike turbellarians and nematodes.

The "tunnel world" sets automatic constraints.[4] "It is unique and quite mind-boggling," explains Gradinger. "Some of the brine channels are just so small that the bigger grazers don't make it in. The physical properties of the brine channel and the network diameters create size restrictions that control the biology." Inside the ice tunnels, it pays to have a very flexible body. Evolution has shaped some of the predators, such as the turbellarians, so that they can squeeze themselves into tunnels less than half their usual body diameter in the hunt for prey.

Watching the inmates of the tunnel world is not easy. The best you can do is to cut a slice of ice about half a centimeter thick, put it under a microscope, and "sit in a cold room at -2°C and see what happens," says Gradinger. "Turbellarians; I like them a lot I have to admit," says Gradinger. "They may look like little red blobs, but under the microscope they are very active. You can actually see them moving around through the brine channels, and see how they squeeze themselves through very, very narrow parts."

Gradinger has found that young polychaete worms are thin enough to squeeze inside some of the channels and graze on the rich pastures within. In the ice off the coast of Alaska, he has seen astonishing numbers of these worms—more than 100,000 of them per square meter—feeding in the ice for a few months in spring. That is a graphic reminder of the abundance of life that ice algae can support.

I worry about what will happen to these unique worlds if the sea ice begins to melt away completely in the middle of the summer. Gradinger explains that the creatures that live in the seasonal ice found close to the shore have always experienced ice that melts in summer and refreezes each winter. In summer, they live in niches among sand grains in the bottom of the shallow seas, and when winter comes they return to the newly forming ice. The richer varieties of creatures that live in old ice far out on the deep Arctic Ocean are different. "These organisms don't survive in seawater or on the seafloor, so if you remove the sea-ice cover you remove these species," says Gradinger bluntly. That means that with the loss of the old sea ice a unique, almost unknown world may vanish. "Yes," says Mundy, "you can think of these animals as like the polar bears, but the polar bears are visible."

The ice algae are not the biggest source of food in the Arctic—plankton blooming in open water are more productive (when and where there is open water)—but they do have a special importance. Ice algae begin growing as soon as the sun returns to the frozen Arctic and light shines through the ice. "The sea ice provides a very early pulse of food, from March to the end of April, that is available for herbivores long before anything is happening in the water," explains Gradinger. "At that time of the year, concentrations of algae in the bottom layer of ice are four orders of magnitude higher than what you see in the water column. It is an enormous, high-density food patch." The ice algae kick-start the ecosystem in spring. The tiny grazers—the amphipods and copepods—that feed under the ice are also the key food for enormous numbers of fish, especially the Arctic cod that is a vital link in the web that feeds birds, bigger fish, seals, and on to bears.

Later in the year, in the parts of the Arctic where the ice melts away in summer, phytoplankton will bloom in the open water where there is far more sunlight than down under the ice. As a result they will provide an even bigger pulse of growth than the ice algae (provided there are enough nutrients in the water). That leads immediately to an interesting question: in the future, when the Arctic loses more and more summer ice, and more open water is exposed for longer each year, will the Arctic become a more productive, if very different, sea than it is now?

The answer seems to be that it will. Already the first signs of change are arriving: after the sudden fall in the sea-ice area in 2007, plankton started to grow in the newly exposed open water over a longer season. In a single year the total productivity of the seas leapt by 40 percent. That change was recorded by Kevin Arrigo, a biological oceanographer from Stanford University in California, who took measurements from a satellite-borne sensor that measures the "greenness" of the surface ocean.[5] I ran into Arrigo while we were both waiting to board a plane in Chicago en route to the American Geophysical Union's giant conference in San Francisco. I asked him what he thought of that huge change. "I was very surprised," he said. "Forty percent is a big increase. I thought there would not be enough nutrients in the water for such an increase. There must be some process bringing nutrients up from the deep water below." He explained that 30 percent of the increase he saw came from the greater area of open water available, but 70 percent came because there was simply a longer time when the seas were free of ice.

Whether the whole Arctic would always be a more productive place if the ice shrank away to nothing every summer is more debatable. The Arctic seas might end up capped by a layer of freshwater from melting ice and faster flowing rivers that prevents nutrients from getting up to the surface. Or the open waters might be stormier and mix nutrients more thoroughly. In the future, says Mundy, "you could have nutrient limitation or you could have open water plus the ability to mix the water better by coupling it to the atmosphere. Which one will take over we don't know yet." The Arctic is a big place and the answer will not be the same everywhere. "You really have to think of the Arctic Ocean as a patchwork of different systems, and each of the systems has its own rules and regulations," says Gradinger.[6]

A leap in the ocean's total productivity might sound like good news for Arctic residents, and it could be if they were all very adaptable to new conditions. But such a radical change could just as easily be a disaster that simply encourages different animals to move into the Arctic. The special creatures that live in the Arctic now are adapted to its particular rhythms and to the timing of the ice melt. If that rhythm changes, then the whole ecosystem may be thrown out of balance, destroying the web of relations within the Arctic from bottom to top, leaving the Arctic a very unfamiliar place.

We have some clues as to what could go wrong from research that is already underway across the Arctic at the University Centre in Svalbard. Here, marine biologists Jørgen Berge and his postdoctoral researcher, Janne Søreide, study blooms of plankton beneath the ice and in the nearby seas. I found out about their project CLEOPATRA (Climate Effects on Planktonic Food Quality and Trophic Transfer) by chance when I saw some pictures from the project in the little museum in Longyearbyen.

One picture showed a group of researchers in fur and woolly hats having a picnic in the snow outside a small, tent-shaped wooden hut with a bent stovepipe sticking out of its roof and a much-patched old wooden door as its entrance. Another showed their laboratory bench, which seemed to be made of planks perched on a row of red, orange, and blue-and-yellow-striped oil drums, parked inside a tent. This looked like fun and I called them up to find that they have a delightful tale to tell, but one that could only be put together by researchers who are happy to live with cold, ice, and bears.

Berge and Søreide work out at Rijpfjorden, a remote spot on the most distant northeasterly island of the archipelago. They have to fly out there by helicopter and live for weeks crammed into that wooden hut I saw. A large tent does indeed serve as a laboratory. "It works very well. It's not well insulated, but we don't need so much comfort," says Søreide stoically. "We could sleep in tents, but as this place is known for being crowded with polar bears, the hut is better." So far the bears have been more interested in chasing seals than researchers. "We have curious bears but not problem bears," says Søreide. Still, everyone obeys the strict laboratory regulations and carries a rifle.

Everywhere in the Arctic, the zone alongside the edge of the sea ice where phytoplankton bloom in spring as the ice melts is immensely important. The fresher, lighter meltwater helps trap the algae near the water surface where there is lots of sunlight and the bloom may stretch thirty miles out across the water. Satellites can easily pick out the band of pale green from space.

The ice-edge zone is always buzzing with life. Seals and larger seabirds are busy catching fish, walrus climb out on the ice in between diving to the shallow bottom to feast on clams, smaller birds like the dovekie are dipping into the water to pick out copepods, piratical jaeger rob less agile birds, whales swim along the ice edge trawling up plankton or catching fish, and polar bears wait patiently at the ice edge in hope of snatching a seal. In a day's cruise along the ice edge to the north of Svalbard, all these scenes will pass by, just as they will elsewhere in the Arctic where the summer ice edge lies in shallow waters. Deeper inside the Arctic, the seas are frozen over and there is a great silence.

Among the abundant life at the ice edge, the cold-water copepod *Calanus glacialis* is a critical link in the food chain right across the shallow Arctic seas. Although just a tiny shrimplike creature, six millimeters from head to tail (excluding its two long antennae), *Calanus glacialis* is the most important grazer of the Arctic seas. It is found in the billions near and under the ice where it fattens up at great speed. For the copepods, "fat-

ten" is the right word. As they gobble up plankton during the short summer, they build a store of high-energy lipids that will see them through
the cold, dark winter when there will be nothing to eat.

To huge numbers of fish and birds, this copepod is a perfect little package of high-energy food. Copepods provide a royal feast for Arctic cod, *Boreogadus saida* (quite different from the larger Atlantic cod that we eat and
called "polar cod" in Europe), which live under and close to the ice.

Out at Rijpfjorden, *Calanus glacialis* live in special abundance, and
Berge and Søreide have found that it has a clever trick to make the most of
the ice and its melt. When the sun returns to the Arctic in spring and rises
high enough for its rays to penetrate through the ice, the algae that live
beneath immediately begin to grow. The huge ice-edge bloom comes later,
when the ice starts to melt. The copepod takes advantage of both blooms.
In mid-April, they feed on the spring ice algae and reproduce. When the
ice-edge bloom comes in July, their offspring feed, develop, and build up
lipid reserves for the winter. "By taking advantage of the earlier ice algae
bloom and then the phytoplankton growth season, *glacialis* can fulfill
their life cycle in one year, despite living in a harsh Arctic environment,"
explains Søreide.

The copepods must have their timing right. Up here in the really high
north the summer is very short. "The time in between these two blooms
may also be critical," says Søreide, "because it takes time for the copepods
to develop so they can feed on the second bloom." In a warmer world, when
the ice melts earlier and faster and the seas are warmer, the timing could
all go horribly wrong

Calanus glacialis has a smaller, skinnier relative called *Calanus finmarchicus* that prefers warmer waters. A little farther south it lives in staggering numbers and feeds the herring and Atlantic cod of Iceland's seas. But
it's not big enough to provide for the creatures of the High Arctic. "The
energy content in these two *Calanus* species is so massively different," explains Berge. "There may be as much as five times more energy in *glacialis*
than in *finmarchicus*."

If the summer ice vanishes quickly in a changing Arctic, so too will
those ice blooms. Because *Calanus glacialis* is geared to cold water and the
timing of the ice blooms, it will be at a disadvantage to its relative *finmar-*

chicus, which is adapted to warmer, open seas. "A change in ice conditions would mean a change in the composition of *Calanus* species, which would have an immense impact on top predators including whales and seabirds," says Berge. Already more southern areas of Svalbard that have had several ice-free summers have been taken over by the southern species, while the more northerly areas are still dominated by the High Arctic *glacialis.*

The story of Rijpfjorden and its copepods may be one tiny tale from one far corner of the Arctic, but it reinforces a bigger point: it is not just the total productivity of the Arctic seas that is important to the web of life it supports but also the "when" and the "where," the timing and the place— leading to what ecologists call "match-mismatch" problems. A tiny cope- pod that is well adapted to the old Arctic gives way to one from the south that does better in the new Arctic. But the whole shape of the pyramid may then change too. How big a change might come, we don't yet know.

In the more southerly Bering Sea, we already know of a major change to the ecosystem that is driven by changes in the "when" of the sea-ice melt. The phytoplankton don't just feed the zooplankton, fish, and birds in the waters around them. A rain of them also sinks slowly to the bottom of the sea, giving life to the shellfish, crabs, and bottom-dwelling fish that are particularly abundant in the Bering Sea. They in turn feed larger animals that dive to the sea bottom in search of prey. Both pelagic (water-dwelling) and benthic (bottom-dwelling) creatures are ultimately dependent on the phytoplankton which capture the sun's rays at the surface. But which set of creatures will get the biggest share of the food partly depends on the timing of the sea-ice bloom.

If the ice edge bloom starts early in the year, the water is still too cold for many of the zooplankton and fish that would usually eat them to ap- pear in large numbers. The result is that fewer plankton are eaten and more fall through the water to provide a bigger feast for life at the sea bot- tom. If the plankton blooms later, then the water is warmer and the zoo- plankton and fish are there to mop everything up. Life in the water gains and the creatures of the sea bottom lose. So the timing of the bloom has a big impact on the overall structure of the ecosystem.

Already in the more southerly Bering Sea, the impact of this change is being felt. The whole ecosystem is shifting in favor of sea dwellers and away from bottom dwellers.[7] The number and distribution of fish, clams, crabs, and even jellyfish are all changing as the pelagic zone beats the benthic zone. As the timing of one event changes, the whole ecosystem shifts direction.

Related changes may come to the High Arctic. "Shifting the sea ice causes different water temperatures, which in turn cause different zooplankton to grow. These might consume everything in the water column and then nothing is left to sink to the seafloor," warns Gradinger. "The implications are huge because we have these benthic feeders living in the Arctic such as the gray whales or the walrus or the bearded seals, and for them a shift to a stronger pelagic component will be disastrous because they can't feed off small copepods. You might have a complete shift in the ecosystem structure."

Can we predict what will happen, I asked Gradinger. "There are so many factors that are relevant and everything is always affected," he sighed. "These interactions are not linear, so we can come up with some ideas about linkages, but we can't make a predictive model right now."

The "where" of the ocean's productivity is going to be just as important, and once again its impact will be just as hard to predict. There is one place, however, where we can expect to see a big, unwelcome change. In the new Arctic, if the edge of the sea ice moves too far out to sea, disaster will strike many colonies of breeding birds.

The dovekie, the kittiwake, the thick-billed murre (Brunnich's guillemot), and the ivory gull all rely heavily on the ice and ice-edge zone for food. They also need to nest where they are safe from their two big predators, the Arctic fox and the glaucous gull. Kittiwakes and murre seek out steep cliffs with ledges close to the sea, dovekie (little auk) search out inaccessible rocky mountainsides where they nest among boulders, while the ivory gull prefers isolated mountain tops surrounded by snow. Even so, during the breeding season, the pickings are rich for fox and gull.

Around the great kittiwake colony at Diskobukta, on the west side of Svalbard's Edgeøya Island, I have watched Arctic foxes poised, waiting for a young kittiwake's first flight. If it does not go smoothly and the kittiwake makes a nearby crash landing, it rarely takes off again. Its broken body ends up flapping helplessly in the mouth of a contented Arctic fox, which will carry it away, perhaps with plans to bury it against leaner winter times. Not far north, the gigantic sea cliffs of Alkefjellet on Spitsbergen are the breeding place for 300,000 thick-billed murre. They face the same threat to their chicks. This time it is the large glaucous gulls that are waiting for an easy meal, their own fat chicks sitting alongside them. A baby murre cannot really fly: it must make a tumbling glide from its nest on the high cliff and try to hit the water. Its father waits down there, calling desperately, and if they are reunited, they will paddle away together on a journey to their southerly wintering grounds. If the baby murre fails to find its father quickly, there will always be a gull ready to snatch it up as a meal for its own young. Sitting by those cliffs I've seen the lucky chicks that tumble from the sky and make it safely to their fathers and the unlucky ones that don't. Just occasionally a chick is snatched away by a glaucous gull but manages to struggle free from its grip in midair and escape to the sea.

For the kittiwake, murre, and many other Arctic breeding birds to survive, they need those inaccessible places where their predators can only pick off the unlucky. But these rare places cannot be too far from the ice edge where parents fly to fetch food for their young. As the ice retreats, we are beginning to see the first evidence of the strain being placed on Arctic breeding birds.

In Svalbard, Berge is studying colonies of dovekie. The dovekie is a tiny black-and-white bird living in enormous numbers across the High Arctic and nesting in colonies which may contain several million birds. Despite its numbers it is as vulnerable as the polar bear and ringed seal. Dovekie feed almost exclusively on crustaceans, especially those fat copepods that are abundant in the seasonal ice zone. When they have a chick in the nest, they must feed within the range from which they can carry back food.

Already Berge has seen that in a couple of warmer years, when the ice was farther away and it was harder to reach the copepods, the dovekie's breeding success fell. But he is not drawing any big conclusions yet because "the window of knowledge is too narrow." Dovekie might breed for twenty years and they must have some good and some bad years or "they would dominate the globe," he says.

The Arctic researcher's curse is that decades of patient study and intimate knowledge of a population are needed to spot a developing trend among the enormous natural ups and downs of the Arctic. There are still far too few "baseline studies" to monitor Arctic change quickly and effectively. That's not too surprising. It takes a rare individual who is willing to take on a study that might last a lifetime, especially as long-term research grants are always a terrible struggle to come by and recognition may have to wait for decades.

George Divoky is one such individual and to him we owe some of the few long-term records we have of the impact of the retreating ice on Arctic birds. For the past thirty years, Divoky has spent his summers on an isolated, low-lying island thirty-five miles northeast of Barrow in Alaska, closely following the good and bad times of a colony of black guillemot, a starling-sized sea bird that comes here to breed. There have been obstacles to Divoky's research at every step. It has been hard to find funding, and Cooper Island is not easy to reach. For many years, Divoky just lived in a tent. More recently friends helped him build a small hut.

For his first twenty-five years no one beyond a small circle of ornithologists much appreciated what he was up to. Then, in 2002, a reporter from the *New York Times* spent two months on Cooper Island with him and published a 12,500-word article about his work.[8] Suddenly he was famous. And rightly so, as his patient studies have allowed him to see the first impacts of climate change in a way that no one else can.

Divoky's black guillemots should not really be there. Black guillemots nest in clefts amid boulders while Cooper Island is flat and featureless. But as Divoky discovered back in 1972, the U.S. Navy had dumped old ammunition cases on the island and the guillemots had found that they made perfect places to nest.

That was a big advantage, Divoky told me. He could lift the lids of the boxes and see how the birds and chicks were getting along. "If they weren't breeding in such a strange area I would never have gone up there," he said. "I have complete access to every nest site and that really makes it for me. I have been to those boulder colonies, too, and getting to the nests is really tough."

I first talked to Divoky when I called him in Seattle, his home when he is not up on his island.[9] Webcasts of him out on Cooper Island show a man in his early sixties with a graying beard, wrapped in multiple layers of baggy clothes, some stained with what look like guillemot droppings, talking intensely to the camera.[10] Sometimes he wears a purple-and-mauve wooly hat pulled down over a pair of round spectacles, and other times a complicated white sunhat with visor, ear flaps, and neck guard. His online image is of an eccentric ornithologist, but when I eventually met him a few months later at a conference in Quebec, he turned out to be a lean, clean-shaven, well-dressed, and urbane storyteller with a droll sense of humor and a profound feel for nature. Seattle and Cooper Island are perhaps two sides of his personality.

Divoky has been ringing the birds that nest at Cooper Island (now around 160 pairs) for so long that he has grown to know them and their individual struggles well. Guillemot male and female couples normally reunite when they return to the nesting ground. If one member of the pair shows up late, however, he or she may find his or her partner has grown impatient and paired with another bird. Divoky can tell tales of bloody battles when an aggrieved male arrives to find his partner already taken.

Sometimes disaster strikes and it is personal. A polar bear comes along and kills a bird that Divoky first ringed seventeen years earlier. Another time, a group of horned puffins kills all the guillemot chicks in a part of the colony. "It is very disheartening when you are measuring those guillemot chicks," he says, "and also just because you get connected to them."

Now his intimate knowledge of the birds is telling him that when the edge of the sea ice retreats far away from the Alaskan shore, it rapidly affects the birds' efforts to raise their young. Parent guillemots fly back to their nest with their catch held crosswise in their bills, making it easy to see what they have caught. Their two chicks stay in the nest for thirty-five days and pile on weight, growing from 35 grams to an almost adult 350 grams. That keeps the parents busy as each chick eats its own weight in fish every day and needs to be fed once an hour.

From 1975, when Divoky started his work, through 1990 there was always pack ice close by and no shortage of fat Arctic cod living just under and around the ice. The chicks kept growing fast. Then the ice began to move around. In the bad years that ran from 2002 to 2004, the ice was so far away that the parents began bringing back other inferior fish that lived in the sea nearer by, including the four-horned sculpin, which contains little nutrition and which chicks hate to eat as its head is covered with sharp horns.

In 2006, the ice came close in again and provided a good year; then in 2007, the summer of the great collapse arrived, and the ice edge pulled 200 miles offshore. That was way beyond the fifteen- to twenty-mile foraging range of a guillemot. By closely watching his birds as they flew back to their ammunition boxes, Divoky was able to compare what happened in the good and bad years. In August 2006, 75 to 80 percent of the fish brought back were the nutritious Arctic cod; the next year it was only 20 percent. When food grows scarce, there is mayhem in the colony. Hungry firstborn alpha chicks start to attack and kill their younger beta siblings. "It is a very straightforward story," Divoky says. "The ice retreats, the prey species changes, and we go from two chicks surviving to just one." In 2003 and 2004 every beta chick died. In the earlier, good, years they all lived.

Divoky is looking out for the next and more troubling change. Early in the summer, he waits for his birds to return to the island. In the winter, they live out among the pack ice north of Barrow. "Every year close to 85 percent of the adults have survived," he explains. "That hasn't changed, which means that the winter pack ice conditions haven't really been impacted by the summer changes. The ice is obviously there all winter and there are also enough Arctic cod under the ice." Now he is looking out for any change in the trend in overwinter survival. That would indicate that the Arctic cod living under the ice are beginning to disappear. "That is a major measure of the Arctic that really isn't easily obtained in any other way," he says. "No one is sampling the underice environment except my guillemots. They are sampling it all winter long."

Knowing that something was going wrong with the Arctic cod beneath the ice would be an important warning that the Arctic ecosystem was breaking down. It reminded me of something that Berge told me: "People tend to talk about the Arctic as very fragile. In one way it is the opposite. The ecosystem is extremely robust because the Arctic species are adapted to cope with very high variability from year to year and extreme conditions. But it is fragile in terms of having relatively few important species. If they disappear they will have immense effects on the rest of the system." The Arctic cod is one of those species. For now, it is still there, although increasingly out of range of Divoky's birds when they need it most.

With so many summers spent on Cooper Island, often alone, Divoky seems to me to have gone far beyond the usual ornithologist's fondness for his birds, to develop a deeper feeling for the vanishing Arctic. More and more polar bears are showing up on his island in late summer, and Divoky explains that he is torn by his concern for himself, for his birds, and for the bears. A polar bear walked up to the back of his cabin where he was sleeping in 2008, he told me. "I heard a snort. I got my shotgun and opened the door and the bear was right in front of me. I shot over the bear's head and it ran off. But it wasn't until later that day that I saw where a bear had come out of the ocean. Suddenly these other prints showed up right next to it. A cub had been hanging on to its mother's neck. I could see where the cub had been waiting down next to some driftwood while the mother came to the cabin to look for food. Of course there wasn't any food and I scared her away. The worst thing about scaring bears away now, is there is no place for them to go. They set off swimming but the ice is too far offshore. A number of times, I've walked down this beach, and I see a totally exhausted bear sleeping; it has come out of the ocean and it is totally sacked out. They stay that way for twelve hours or so. I don't want to wake that bear up to scare it away. But if I don't try to scare them away from the cabin I take risks. I also feel awful when I do it."

At the end of this story Divoky pauses for a moment. Then he says, "You know, there are people who say that if you go out to look for the signs of climate change that's what you'll find, because that's what you want to find. But I would give anything for the ice to come back." That is how I feel too, but we both know that it is too late.

Chapter Eleven

INVADERS FROM THE SOUTH

K eep going for another fifty miles or so past the top of Scotland, and you reach the Shetland Islands, the very northernmost part of Britain. At a little over 60° north the islands are at the same latitude as Anchorage, Alaska, and well to the north of Churchill, home to Hudson Bay's famous polar bears. We are still quite a way from the Arctic, but the great sea cliff breeding colonies are packed with thousands of Arctic birds—skua, tern, murre, kittiwake, puffin, and fulmar—and many other animals pass by on their seasonal migrations to and from the north.

I've been visiting these colonies every year for more than a decade, and I've started to notice a host of small changes that might be the signs of a shifting climate. The most dramatic is the much more regular sighting of killer whales close to Shetland shores. Killer whales were rarely heard of here a decade ago. As there is no name for the killer whales in the local dialect—which is much influenced by Norse—but a name for everything else, the killer whales may never have come here in earlier times. Now there are five or so pods of whales up here every year.

At first, each sighting produced a report in the local newspaper and people rushed off to see them. The paper was sternly critical of boats that went so close that they might scare away what could be a big tourist attraction. Then the killer whales' image began to change. Seal colonies around the island seemed to be vanishing, although the wildlife biologists demanded hard evidence before pronouncing the whales guilty. A full-grown killer whale needs to eat a seal, or its equivalent, every day or two so a few missing seals would not have been too surprising.

Then the whales were seen massacring a flock of dunter (as the elegant black-and-white eider duck is known in the local dialect) along with their newborn chicks in 2007. They ate forty of them in a few minutes. The following year the whales showed up at a popular bay and, in plain view of tourists, began not only killing seals but tossing a bleeding, half-dead seal pup back and forth out of the water. Seal tossing is a well-known sport of killer whales that live in the Antarctic but it was not entirely what tourists had been expecting.

Killer whales are spreading through the Atlantic and traveling deeper into the Arctic earlier in the year. Their conspicuous presence farther north is a powerful sign of another kind of change in the Arctic. We've seen how the disappearing ice may harm the polar bear, the ringed seal, and other creatures dependent on ice, and how changes at the bottom of the food chain may transform the structure of Arctic ecosystems. Now we encounter a third kind of change: invasion. As the seas warm and the ice melts, the open seas provide new opportunities for highly competitive invaders from the south.

For the residents of the Arctic seas, an invasion by killer whales is about as welcome as the Vikings were when they first showed up in the Shetland Islands, slaughtering and enslaving the Celtic population. Up in Svalbard, 1,200 miles north of Shetland, Kit Kovacs of the Norwegian Polar Institute has been seeing killer whales swim right up to the edge of the ice in summer. They are arriving earlier and earlier. "This year [2008], we were up at 80° north [latitude] in March and April, and we had killer whales," she explains. "That was really surprising to us because normally you expect them in maybe the months of June and July, when the minke whales are active. It seems that it is part of the new Arctic that these big predators are here very early."

On the other side of the Arctic, killer whales are on the move north too, says Steven Ferguson at the University of Manitoba. He has looked back at old whalers' records and sea-ice data and found that early in the last century, there was ice in Hudson Strait, leading into Hudson Bay.[1] When the ice went, around 1940, more killer whales began to appear. The pattern seems to be continuing. "We think that this probably is going to cascade across the Arctic as you start to lose sea ice. New areas will open up to temperate invasive species like killer whales," says Ferguson.

The new Arctic is looking a little scary. Killer whales up in Svalbard don't just eat seals but will hunt minke whales in a pack. Over in Canada, Inuit hunters report that the killer whales they have seen are after big mammals, not fish.

I've followed killer whales hunting a minke whale in Antarctica's Weddell Sea, weaving around ice floes and flat-topped tabular icebergs in a fast inflatable boat. The hunt was an awe-inspiring and terrifying sight.

The tall, swept-back triangular fins of the killer whales break surface together. In the center of the pack of ten or so of the killer whales was one animal with a smaller, sweptback fin. That was the minke whale, already bleeding where chunks of blubber had been torn from it. Crowds of noisy seabirds followed the hunt, feasting on bits of whale floating in the sea. A decapitated seal—killed by the whales in passing—rose to the surface. Every time the desperate minke came up for air, the killer whales pressed around it to force it back down. They must tire it out for the final kill. On a sprint, the killer can outrun the minke. In a really long distance swim, the minke can outlast the killer.

The killer pack had to make sure the minke couldn't break away. This was a contest among leviathans. The male killer whales are thirty feet long and weigh six tons, while the minke is slightly bigger and heavier. Every animal is the size of a bus and they are going at twenty knots under the water. Again and again the minke desperately surfaced but was never able to shake off the pack that surrounded it. Eventually the chase headed farther out to sea, leaving us behind. We'd already run for more than eleven miles in our little boat and it was dangerous to go farther. I will never know how that particular minke whale fared. From the size of the pack surrounding it, I don't think it had much chance.

Up in the new Arctic, the appearance of killer whales has already terrified beluga. In Svalbard, Kovacs tells me that she has been trying to record their underwater sounds. Beluga whales are very social and are known as "the canaries of the sea," she says, constantly chirping to one another. In her part of the Arctic they are "phenomenally silent." Here, she says, "they only seem to make noise in two conditions, when mother and calf get separated or are in some way stressed. Then they are highly vocal. That's not too surprising. The other time is when a big group forms, not a little traveling group of twenty or even a hundred, but a group of many hundreds. When they get together like that, they are vocal."

Why are the beluga whales silent except when they have great need or are in the safety of large numbers? The answer is easy to guess. It is a classic reaction to the presence of killer whales. When hunting, the killer pod listens carefully for prey but remains quiet to give no warning of their presence: only when they attack is there a sudden blizzard of underwater communication sounds. Then it is too late. The next underwater sound is that of crunching bones.

Even bowhead whales may not be able to cope with killer whales. "They are quite large so you would think they'd be relatively safe from predation, but they seem to rely on sea ice as a means of reducing the risk of killer whale predation. That concerns me," says Ferguson. To find out what happens when killers confront bowhead, he has been turning to the traditional knowledge of Inuit hunters, traveling around Inuit communities and collecting their stories and observations. A scientist might wait years to be at the right spot to see the two giant mammals confront one another, but Inuit have been out there observing nature and passing down stories for generations. "Hunters have pretty graphic descriptions of how the killer whales take the bowhead. Bowhead will stand and fight. They are not fast, but they are quite maneuverable," says Ferguson. But it does not always go one way. "We have even an observation from hunters of a bowhead whale killing a killer whale in a confrontation." From the scars in the tail flukes of bowheads, Ferguson thinks that killer whales often just try to harass the mother to slow her down while they are really after her smaller calf.

The one advantage of all three of the ice-adapted whales is that they can more easily hide in very dense ice than can the killer with its enormous triangular fin. If the ice goes and killers arrive, Ferguson fears a "trophic cascade." That happens when key species are taken out of a food web and the entire ecosystem structure begins to change uncontrollably.

I ask Kovacs if there are other invaders we might expect to see. "You are going to have all of the big traveling animals. Anything that flies or swims is an efficient migrator and an efficient disperser," replies Kovacs. "If you are traveling on air or in water, distance really isn't a big deal. The cost of locomotion is so low. And for a large swimming whale or seal, distance is almost a nonissue. So they are going to move wherever there is food available for them. You have a lot of the large baleen whales moving north and now we are seeing blue whales in aberrantly high numbers in the last couple of summers, although that still means only between six and ten animals. But still, we are seeing them for the first time."

I'd seen a group of fin whales, with their long, sleek bodies and distinctive blow which clouds the air a few seconds before their dorsal fin breaks surface, moving up the east coast of Svalbard. That surprised me as these baleen whales' usual home is much farther south. Kovacs confirmed that there are many fin whales around Spitsbergen now. "They are already on the move and they will take food that certainly could have been taken only by the bowhead whales just a decade or two ago. If they are up there and take in a large volume of food as these large animals tend to do, then you will have pressure on the Arctic dynamics," she warns.

Competition between residents and invaders comes partly down to life strategy, Kovacs tells me. Many Arctic species have "a slow and steady" life strategy. The Arctic is a place of great extremes and great variation year on year, with good years and bad years. One way to deal with that is to live for a long time and just breed whenever you can; the good years and the bad years will average out and in the end you'll leave offspring to take your place.

"The long-term winning strategy is to just do it slow and do it well," explains Kovacs. "So you have only one baby every five or seven years in the case of a bowhead whale, but you live for two hundred years so you get there eventually. All these Arctic mammals are slow to get to sexual maturity but they live long." The same is true for many birds of the Arctic. The tiny, little auk just lays one egg each year when able but lives for twenty or more years. It may not breed successfully for a few years but that is all part of the long-term success strategy.

This conservative strategy works well in the old Arctic, but what happens when conditions improve and you are up against competitors that breed faster? "Slow and steady doesn't work when you have a baleen whale coming up that can have a baby every year," warns Kovacs. "Then there are more of them and they eat large volumes of food. That food would either have fed your young or would have been your food. The Arctic endemics aren't going to be very good at keeping up with those other guys."

When I had been thinking about the future of the bowhead whale in the changing Arctic, my mind had focused on threats to its food supply of rich copepods along the ice edge as the ice vanished. I had not thought that the real threat might be competition from other whales coming up from the south.

All over the Arctic invaders are pouring in. "Harbor porpoises have been seen in Spitsbergen for the last four years," Kovacs tells me. "This has never happened before in history." And on the other side of the Arctic, on Cooper Island off the north coast of Alaska, ornithologist George Divoky started to see horned puffins, which have no affinity with ice, arrive and then become regular breeders over the last decade. Now another puffin has shown up. "In the last couple of years, I have seen a few tufted puffins. That's a true southern species that hasn't been here before," he told me.

What about the less conspicuous creatures? Creatures that don't swim or fly are just drifting to the Arctic if the currents head that way. "Blue mussels are back for the first time in a thousand years," says Jørgen Berge of the University Centre in Svalbard.[2] Blue mussels, a staple of Spanish and Portuguese cuisine, were common in Svalbard during the Viking age, when northern Europe was in the middle of the medieval warm period, but they had long since vanished. Now they have returned.

Other creatures that are happy in warmer waters are also showing up. "We talked to fishermen who have been here for the past thirty years, and they hardly found any cod in the '70s and '80s," says Berge. "We started to see some cod in the trawl hauls in late '90s, now there is a really huge amount of big Atlantic cod. We see larvae of haddock we hadn't seen before. Since we are living in the area where the main inlet of the water from the Atlantic Ocean comes into the Arctic, carrying huge amounts of eggs and larvae from southern species, there will always be a massive influx of potential new settlers. What has changed is that some seem able to settle and grow."

As southern invaders enter the Arctic, indigenous species might be expected to flee north into the cold Arctic sea, but as Kovacs points out, there is really nowhere to go. "You can't just retreat northwards, because you fall off that coastal shelf into the deep Arctic Ocean." Out in the deep, the seas are not productive and there is not enough to eat.

As the boundary between the sub-Arctic and the Arctic moves north, and the seas warm, not all the invaders will be large or obvious. Up in the Arctic, it is too cold for many disease-causing bacteria and viruses to thrive. "The Arctic animals are really very much protected by cold. In a warmer environment, you will have pressure from introduced diseases and from introduced parasites. These potentially disease-carrying animals will thrive in a way they never have before," worries Kovacs. "If you get a cut when you are working with the Arctic animals, you never worry about it, because they don't carry anything. Whereas working in the southerly latitude, my God, you cut yourself when you are doing something with seals, you have to detox yourself, because you can end up in the hospital with blood poisoning. It is a totally different world." It is a world that is coming to the Arctic.

The deep northern seas will not be the final frontier for Arctic invaders. There is an even bigger different dimension to this mass movement of species that I had completely missed. The Arctic seas are not just a place that you can move into but are also a shortcut between the Pacific and the Atlantic oceans. I knew nothing of this possibility until I came across a paper[3] published in 2008 in *Science* titled "The Coming Arctic Invasion." Two scientists warned that as the Arctic ice vanishes, "an even more dramatic interoceanic invasion will ensue." Geerat Vermeij of the University of California at Davis and Peter Roopnarine of the California Academy of Sciences in San Francisco predict that many creatures from the Pacific Ocean will enter the Arctic and go on to successfully invade the Atlantic, transforming that ocean as well. The warming of the Arctic thus opens the gates for a transformation of the world's oceans.

"Dramatic" is definitely the right word for this interoceanic invasion. Why do these two scientists think it is going to happen? Not for the reasons you might expect. They haven't been out collecting new creatures in the Arctic. Instead, they have looked at the great trends that have transformed the earth over many millions of years and from them they have predicted the future.

Vermeij and Roopnarine both study ancient mollusks, which with their hard shells are easily preserved as fossils in layers of sediment laid down over millions of years. The two scientists say that the fossil record shows that creatures from the Pacific began to invade the Atlantic via the Arctic about 3.5 million years ago when the earth was warmer. The invasion ground to a halt with the arrival of the ice age when the Arctic froze up. Now with conditions warming again, the invasion can resume.

I'd never heard of anyone making predictions about the near future based on events that happened millions of years ago, so I called Peter Roopnarine in his laboratory at the California Academy of Sciences and asked if this was the usual practice in paleontology. He laughed gently at my question. "No, not regarding climate change now," he said, "but in the fossil record we often have periods of climate change. We've had many of them over millions of years, and we've also had barriers, whether of climate, oceans, or physical, being removed over time, so we have studied the invasions and migrations of the past."

To think like a paleontologist, you have to stretch your notion of time. "During the later Miocene era, just 10 million years ago, we had a climate optimum on the planet where conditions were very mild," Roopnarine says, "then as we moved into the Pliocene, the planet became quite warm. During this warming, starting about 3.5 million years ago, species moved from the North Pacific to the North Atlantic, but by about 2 to 2.5 million years ago, we had a refrigeration of the northern hemisphere and migrations came to a more or less abrupt halt."

The imprint of those events of 3 million years ago can be seen in the Atlantic where descendants of the creatures that crossed from the Pacific still remain. Now, says Roopnarine, there are seventy-seven species with the potential to invade the Atlantic from the Pacific. And that is just the mollusks. But why should Pacific species be able to overrun the Atlantic?

The answer turns out to be a familiar story of puny Europeans not being able to take on bigger, more competitive Americans. At one time, the Pacific and Atlantic fauna were equivalent, explains Roopnarine, but over many millions of years, the rich, upwelling waters of the Pacific have promoted intense competition and the evolution of bigger species. In the Atlantic, by contrast, there has been a series of extinctions. "If you pick any guild of mollusks," says Roopnarine, "whether they are suspension feeders or grazers, they are larger in the Pacific and competition in the Atlantic will be dominated by these Pacific species."

The barrier that holds back this invasion of bigger beasts is not the cold of the Arctic, but its lack of food. Once the ice melts in summer, the seas will receive more sunlight and more plankton will grow in the warmer waters, just as we are beginning to see. The "food barrier" will fall. How long will the invasion take then? "When conditions are right, and once they begin to move, it is a matter of just a few years before they are fairly widespread," says Roopnarine.

Does the same argument apply to many other creatures of the Pacific seas? "I think it would be safe to say that other organisms that have not been documented geologically but are ecologically similar are probably going to do the same thing. We have a lot of echinoderms [starfish], coelenterates [jellyfish and anemones], and fish ready to move." By the 1990s, plankton that were once specific to the Pacific had begun to move to the North Atlantic.[4]

The travels of starfish, sea anemones, and plankton may pass unnoticed by all but marine biologists, but fish are a very different matter. We know that fish can move rapidly when waters warm and whether they enter the Arctic from the Pacific or the Atlantic, fishermen and the big industry behind them will take note.

Two areas of the Arctic and its adjoining waters, the Bering Sea and the Barents Sea, are among the world's richest fishing areas. Over half the fish harvested in the United States come from the Bering Sea, while more than half the fish eaten in Europe come from its Arctic waters. A lot of people—fisherman, fishery regulators, and pirate fishermen—are watching out for changes in the Arctic's fish with dollar signs in their eyes.

Change can come with great speed. I visited a small museum in the town of Ilulissat, which is Greenland's biggest fishing port. There I was told the strange tale of how a small shift in the temperature of the nearby seas led Greenland's indigenous folk to dance the foxtrot, and even the tango, a remarkable case of how fish can affect human behavior.

In the 1920s, a branch of the Gulf Stream, the Irminger Current, changed its pattern of flow and brought warmer than usual water to the west coast. Seals retreated north and Atlantic cod flooded in. All of a sudden, seals, the mainstay of Greenland's indigenous hunt, were hard to find, but cod were plentiful and easy to catch and sell on the international market. Cod fishing ventures and salting houses sprang up at great speed all over west Greenland.

That brought cash to the indigenous economy, and glossy mail-order catalogues from Denmark were not far behind. Sealskin boots went out of fashion as Western styles arrived along with new furniture, foods, and—most desirable of all—the gramophone and popular music of the day. Out went the traditional music of song and drum, along with the reels and foot-stomping dance tunes brought to northern communities long ago by Scottish whalers. Young Inuit took to the tango and foxtrot.

The good times eventually went too far. In the early 1960s, 400,000 tonnes of cod were being caught per year, which was more than the seas could bear. Overfishing and a more gradual temperature change finished off the cod. Shrimp fishing largely replaced it and that is now Ilulissat's biggest catch.

The shift in the Irminger Current is a particularly famous example, but it is only one of a great many natural changes in the seas that have sent fish swimming somewhere new. Climate clearly matters. Nevertheless, I was not surprised to hear that an exhaustive study of the future of Arctic fisheries commissioned by the Arctic Council concluded that: "The total effect of climate change on fish stocks is probably going to be of less importance than the effects of fisheries policy and their enforcement."

It is easy to see why. Almost everywhere you look, greed and the inability of governments to set and enforce sensible rules are destroying the world's fish. At the peak, in 1968, factory trawlers from all over the world were taking 800,000 tonnes of cod a year from Newfoundland's Grand Banks, a fishery so rich that early visitors claimed you could catch cod simply by lowering a basket over the side of the boat.

By 1992, the Grand Banks' wealth was utterly exhausted: the fish were gone, the fishery closed, and 30,000 jobs were lost. The cod have never come back. Around the Atlantic, you can name almost any fish you like and then find a date when stocks collapsed owing to overfishing. Take the Norwegian herring, for example: 2.5 million tonnes were taken in 1965 and a total collapse followed in 1972. Capelin fishing replaced herring fishing. That collapsed in the mid-1980s. Almost everywhere too many fishermen chase too few fish, and national governments cannot agree on the right action, or dare not implement the measures that they know are needed.

On the Pacific side of the Arctic, the fisheries managed by the United States have fared much better. Their part of the Bering Sea has the reputation of being the best-managed fishery in the world, and its catch of walleye pollock is enormous. But take a step west into the "Donut Hole," an area just outside U.S. waters and east of Russian waters, and we return to the usual tragedy of the commons. At its peak, in 1989, trawlers from Russia, Japan, China, Poland, Korea, Spain, and other nations rushed to its unregulated waters. In 1992, the fishery collapsed. It has not recovered. On the Atlantic side of the Arctic, a similar story can be told of the "Loop Hole." It is an odd patch of international water, hemmed in on all sides by the 200-mile exclusive economic zones that extend from Russian or Norwegian territory. In the early 1990s, changes in ocean temperatures started bringing large numbers of cod to the Loop Hole.[5] Norway and Russia wanted to manage this new fishing ground in a sustainable way, but fishermen from other nations, especially Iceland, had already shown up and were taking all the cod they could. Only after nine years of talks was an agreement reached. By then the Loop Hole had been fished to exhaustion.

These stories of fishermen and the failure to control them have a special significance for the new Arctic. Commercial fish from southern waters are moving north. There are the Atlantic cod moving up past Svalbard. Over in the Bering Sea, pollock are moving north and bringing with them the salmon that feed on them. Juvenile pink salmon are showing up in the rivers that drain into the Arctic where they have not been seen before. This is the beginning of the new Arctic ecosystem that is forming as the old Arctic dies. If the new Arctic is ravaged by uncontrolled fishing, if invading fish are followed by invading fishermen before a new marine ecosystem can establish itself, we could end up with the worst of all possible worlds, neither a new Arctic nor an old Arctic, but an Arctic desert. There is a sad joke among fishery scientists that the Arctic could end as a "jellyworld" in which only jellyfish, a form of life that needs little energy to sustain it, is left wandering in the open Arctic seas. A more likely end is that only very fast reproducing species like squid and shrimp would survive.

The risk that the new Arctic could be destroyed before it is born has been quickly understood in Alaska and for once we can be upbeat. The Arctic seas that fall within U.S. boundaries were closed to commercial fishing in January 2009, thanks to pressure from the Juneau-based Marine Conservation Alliance. Curiously, that alliance was set up by commercial fishermen, not environmentalists. David Benton, the alliance's director, who once represented Alaska in negotiations with Russia over the Donut Hole, explained. "We suspect that the fish stocks are moving north. And we also suspect that even some stocks already in this far north are becoming more accessible, as the sea ice retreats. As an industry, we don't want to be part of the problem. So let's prohibit fishing up there right now in our own waters and get a good scientific picture of what's happening in the environment up there before we open up, if ever."

The decision to close around 150,000 square nautical miles of sea came on top of a U.S. Senate resolution directing the State Department to open negotiations for an international agreement on Arctic fish stocks. "We think that there is a much larger conservation problem looming both on the Canadian side and on the Russian side and international waters in the Arctic basin," Benton explained. How fast might we see progress, I asked? "Well, quick in diplomatic time, slightly less slow than glacial time," Benton replied with a laugh. The good news is that talks are going ahead.

Even if new fisheries can be well regulated, the Arctic will still be left with changes to its fish life that may have profound consequences on its mammal and bird life. Arctic cod (*Boreogadus saida*) is a keystone species that is not taken by commercial fishermen. It weighs only three or four ounces, with a thin body that tapers away sharply from a large head with big eyes. Seals and birds rely upon Arctic cod to fatten themselves up and the cod in turn relies on amphipods and copepods, like the lipid-filled *Calanus glacialis*. Much about its life is mysterious. "In the summertime they are dispersed all over the place so they are very difficult to find," says Louis Fortier, a professor at Université Laval who studies fish and zooplankton population dynamics. "But in the winter time they start to aggregate on the shelf seas some 150 meters to 300 meters down. Then you can see they are really abundant." Fishery scientists using an acoustic scanner found 900 million Arctic cod with a total biomass of over 30,000 tons in just one small area of the Canadian Arctic.[6]

Fortier has been watching seals swim down to feed on Arctic cod. The seals have almost become "research collaborators," he says, as he had up to ten of them living in the moon pool (a circular pool in the middle of the ship giving access to the sea beneath) of his research ship. They grow extremely fat in the winter time. "Sometimes they come into the moon pool, and they would be so stuffed with Arctic cod that they would regurgitate a few specimens that we would quickly catch up and include with our own sampling," he says.

Arctic cod are well adapted to ice and live under it, perhaps partly to avoid being eaten by birds. Nevertheless, Fortier points out, if there is less ice, the species will probably do better at first. "If you have less ice, you have more primary production and there would be more copepods and those small crustaceans to feed on," he says. But then the competitors who have been kept out because they aren't so well adapted to ice will arrive. "If conditions improve too much for Arctic cod, then they will become suitable for other species, especially capelin (a small smelt) and maybe the sand lance. They will outcompete the Arctic cod. When the ice disappears, Arctic cod are going to be replaced by other species."

At the same time, the large lipid-filled copepod species that the Arctic cod like to eat may gradually be displaced by more southerly species, as we have heard. Then the Arctic cod will have more problems. Arctic cod will not disappear, says Fortier, but by midcentury its huge population might become much thinner. As ringed seals are extremely dependent on them, they too will have problems. "Whether the ringed seals will be able to adapt to the reduction in its main staple food and shift to other species we don't know," says Fortier. And of course, the polar bear is in turn dependent on the ringed seal.

The Arctic is seeing a more dramatic change to its environment and ecosystems than any part of the planet has seen for many thousands of years. The Arctic is turning into a new kind of sea; frozen over in winter with a thin layer of new ice which melts way in summer to open water. The seas will be free from the thousands of square miles of old hard ice, rich with life within it, beneath it, and on top of it, that the old Arctic knew. In the warmer open seas plankton may bloom—we can't be sure how richly—but it will be a different kind of plankton, blooming in different places and at different times, feeding the more southerly species that are already adapted to the seas that the Arctic will come to resemble.

As this great change happens, a symbol for the new Arctic looks set to emerge. The reign of the polar bear is coming to an end. We don't have to assume that the polar bear will disappear. It is unlikely to move onto land because it would run into competition with the grizzly bear and lose. More likely it will hold out, at least until late this century, at the top of Ellesmere Island and Greenland where the currents push ice up against the land.[7] "I think the polar bears and some of these ice-adapted whales and animals will still probably be able to survive there," says Ferguson. "From a management and conservation perspective, it's going to be difficult," he says, "because the actual number will be low. But they might survive."

While the polar bear lingers on in a tiny part of his former kingdom, the next ruler of the Arctic will be roaming the seas. The new Arctic and its open summer waters will belong to the killer whale.

OIL & SHIPS

Chapter Twelve

THE BATTLE FOR ARCTIC OIL

Alaska was the first place that I went to figure out where the world's oil companies would make a push into the Arctic seas. I already knew that the U.S. Geological Survey had excited the oil industry by describing the Arctic as "the largest unexplored prospective area for petroleum remaining on earth," with 13 percent of the world's oil and 30 percent of its gas.[1] With a third of that oil in Alaska's offshore waters and with a huge and highly developed Arctic oil industry sitting there on the North Slope, Alaska seemed the place to look for action.

I quickly discovered that it was not that simple. In late 2007, I found a battle was under way. On one side was the oil giant Shell, which had been planning to drill exploratory wells seventeen miles off the north coast of Alaska in the Beaufort Sea that summer. It had an impressive armada. There was the circular, cone-shaped Arctic drill ship *Kulluk*, designed to shrug off any ice that came its way, the high-tech *Frontier Discoverer* drill ship with a derrick taller than the Statue of Liberty, icebreakers to escort them, support vessels, and a truly impressive oil spill emergency response flotilla, including a special vessel, the *Nanuq*, painted blue and white on the advice of local hunters so that it would blend well into the environment.

On the other side were many residents of the community of Barrow, home to the North Slope's most famous native whale hunt and led by its mayor, Ed Itta, along with environmental groups, including Pacific Environment, the Center for Biological Diversity, and the Sierra Club.

Shell said that they had held innumerable meetings around Alaska and talked to local residents. They had come up with a conflict avoidance agreement with the Alaska Eskimo Whaling Commission, which meant

Shell would stop work when the bowhead whales arrived so that the traditional Inupiat whale hunt could go ahead without disturbance. Over a million dollars had gone to acoustic research on the underwater calls of whales.

Still, Shell's efforts were not enough. A lawsuit had begun, not against Shell, but over whether Shell had been granted proper approval by the Alaskan Minerals Management Service for the project, given that the noise of drilling might upset the migrating bowhead whales. The suit had brought the project to a halt. When I arrived, the entire fleet of drill ships, support ships, and spill ships was just standing by in the Beaufort Sea at unbelievable expense.

In Anchorage, Patricia Cochran, chair of the Inuit Circumpolar Council and a native of Alaska, explained to me that oil had brought benefits to people living on the North Slope, including "the school system, the children being able to be taught in their own language, and the infrastructure." But she explained that, "we also have to think of the impacts on our culture. People just adamantly do not support offshore oil development. We have been pretty strong in saying that this is as far as it can go. To our Arctic communities, whaling is such a part of our culture that the loss of whaling would be devastating."

The hunt is still carried out from small boats with a man at the bow carrying a harpoon that he must thrust deep into a spot just behind the whale's head. Nowadays a successful strike triggers an explosive charge carried on the harpoon which kills the whale instantly. Then the whale must be pulled to shore. Crowds gather the moment that there is news of a kill and acclaim goes to the whaling captain taking the first whale of the year. The whale's meat is shared among family, friends, local people, and the wider Inupiat community throughout Alaska.[2] Taking a whale, or as the Inupiat put it, "going out in the hope that a whale will give itself to you," feeds an entire community and reinforces all the values that hold it together. The worry is that even small disturbances to the environment might change the spring and fall migration routes that bring the whales past Barrow and put them beyond the whalers' reach.

The next day I heard Shell's head of Alaskan operations, Rick Fox, speak at an oil conference.[3] Fox outlined the Alaskan Minerals Management Service's estimates of the riches that may lie offshore in the state: "25 billion barrels of oil and 120 trillion cubic feet of gas." That is a serious amount, given that the oil comes close to a tenth of Saudi Arabia's claimed reserves of 260 billion barrels of oil, and the gas is 2 percent of known world reserves. He explained Shell's commitment to the "new frontier," adding confidently, "At Shell, we are focused on the Arctic, specifically the offshore. We know the world is watching us and frankly we don't mind the attention. We came expecting to earn our right to be here every day." Fox listed the consultations that had been held, the exceptional quality of its oil spill response team, the thousands of jobs that would come to Alaska, the hundreds of millions of dollars in contracts over the long life of offshore work and the money that would go into the state's coffers. He explained U.S. energy needs and national security concerns and that "we hear a lot about alternative fuels and Shell is involved in that, but for the foreseeable future fossil fuels, especially oil and gas, remain the bread and butter at the heart of our system." He signed off with Shell's commitment to Alaska and to finding "a common ground."

On the oil side, there's money, jobs, energy needs, national security, and the need to sustain the pipeline that keeps the money flowing into state coffers before it is too late. On the other side are the whalers, and while they don't have the big bucks, they do have the passion. A couple of months later I met a couple of North Slope leaders at the Arctic Frontiers conference in the north of Norway. Robert Thompson, an Inupiaq Eskimo wildlife guide and hunter, is one of the founders of REDOIL (Resisting Environmental Destruction on Indigenous Lands), which is "dedicated to preserving the land in order to preserve our culture."

A quiet man, he was at first suspicious of my questions. "I have lived in Kaktovik twenty years, before that in Lake Minchumina, made my living fur trapping—marten and beaver mostly. The oil industry, I don't see them as nice people, they are a for-profit corporation, they don't act as if they have a soul or if they care," he said. What about all that oil money that would flow from offshore? "Yes," he said, "85 percent of the economy of Alaska, maybe a lot more, is from oil, but the fact that we benefited because they are pumping our oil doesn't mean I am going to sell my body and soul. There is a saying," he continued. "You can either be an oiler or a whaler; it is money or *muktuk* [the tastiest part of the whale], we've got to choose, if you want to keep the whaling culture we've got to protect the whales."

George Edwardson was there too. A big man, his hair tied back in a ponytail, he is the president of the Inupiat Community of the Arctic Slope, which encompasses nine communities including Kaktovik and Barrow. He is full of stories, and he reckons that 80 percent of the people up there are related to him, "within third cousin. We are not so much a race or a tribe as a family. Our problem," he says, "is that there is no baseline study that has assessed the environment properly, without it we cannot see the damage that is going on to the environment there. We have had a vote. Our stand has been very well defined. Each community was called by their names and all said 'No' to any kind of offshore development."

As you might guess, the search for common ground did not succeed. After leaving Alaska I continued to call friends and contacts I had made there. The summer of 2008 passed with lawsuits continuing and Shell unable to get on with its work. This time it decided to pull back its fleet and send its contractors home until the legal issues were cleared. The community in Barrow also made a choice. In October 2008 they reelected Mayor Ed Itta, who had supported the lawsuit against Shell with the slogan "our ocean is not for sale." At first, it looked as though they had backed the winning side. In November 2008, the court ruled that the Minerals Management Service had illegally approved Shell's drilling plans as they had provided insufficient evidence that it would cause no harm to the migrating bowheads. Shell cancelled any plans for drilling in 2009. Indigenous and environmental groups were thrilled: a full and lengthy environmental survey would surely now be needed. But in disputes over oil it never pays to celebrate too early. A few months later, in March 2009, the court suddenly vacated its ruling, which meant that it intended to consider the case all over again.

The same month an economic analysis of future offshore oil and gas development commissioned by Shell from the consulting company Northern Economics of Anchorage was released.[4] The analysis looked out over the next fifty years and envisaged seven production platforms in the Beaufort Sea (oil by 2019, gas by 2029) and four in the Chukchi Sea (oil by

2022, gas by 2036) with fifteen to twenty-four years of exploratory drilling. That activity would create 35,000 new jobs and $15 billion in state
revenues (with oil at $65 a barrel).

Oil is a very long game and fifty years of benefits cannot be dismissed
lightly. Many oil companies had already bet that the lawsuits would eventually go their way. In the February 2008 sale of oil-prospecting leases
for the Chukchi Sea, $3.4 billion was paid for 2.76 million acres of off-
shore leases. Among the buyers were StatoilHydro of Norway, which was
anxious to apply its skills learned in Arctic Norway; Eni of Italy; and
Shell—which spent a staggering $2.1 billion—along with U.S. companies.
Another lease sale, this time for the Beaufort, was planned.

It wasn't long before those bets looked a little less certain. In April
2009, a Washington, D.C., court ruled that the entire 2007–2012 Outer
Continental Shelf leasing program—of which the new Chukchi Sea leases
had been a part—had been approved without adequate review of its impacts and should be reexamined. That meant that new lease sales would
stop, leaving a big question mark over whether the leases that had already
been sold were still valid, and triggering more court action. Meanwhile,
Shell decided to come back with a new scaled-down program for the Beaufort Sea (which had been leased to them in the earlier 2002–2007 program) and leave the ongoing court case behind.

At this point in the story, the question is whether this great legal
game, which is sure to have many more twists and turns, will ever come
to a definitive conclusion. With every passing year, as the ice disappears,
the stress on Alaska's polar bears, bowhead whales, seals, and fishing industry is sure to grow. The additional impact from the noise of seismic
exploration, drilling, and possible oil spills becomes harder to justify. The
pressure to move away from fossil fuels will grow too as climate change
continues. But on the other hand, look at the benefits to the Alaskan economy and U.S. energy security. And energy alternatives aren't ready yet. If
oil prices rise steeply in the long term, as many predict, then the stakes in
this game are going to grow. No conclusion is certain.

I asked Whit Sheard—Alaska program director at Pacific Environment, one of several environmental groups that are leading legal actions
to stop offshore drilling—for his view. "I think we will still be here in the

next couple of years fighting this," he replied. "Whether or not the United States really does put forward a commitment to renewables and conservation will partly determine what happens. Whether or not we have the political will to really take this on as if it were our race to get a man on the moon I don't know."

The State of Alaska is playing it safe. It's pushing hard for a gas pipeline from the North Slope that will connect to the North American grid. There are lots of stranded gas fields onshore which are ready to be exploited, but there has never been a way to take the gas down south. If the pipeline is built, Alaska will have a second revenue stream if the oil slows. Oil may find a new lease on life in Alaska without going sixty miles offshore too. There's less objection to taking near-shore oil by sophisticated "long-reach" drilling from established sites. British Petroleum is planning to break new records by drilling out almost horizontally for over seven miles to reach oil in the Liberty Field off the coast.

I left Alaska no longer willing to place any bets on the United States charging into the Arctic seas any time soon, but also sure that neither oil companies nor environmentalists would quit the struggle. The immediate action looks more likely to happen in Russia, the clear winner in total hydrocarbon resources, according to the USGS survey. Alaska did come first for proven oil resources, but Russia has staggering quantities of gas as well as oil in the Arctic. Onshore, development in Arctic Russia is already roaring ahead and the political picture is very different: there is uninhibited enthusiasm for more. Russia knows all about the horrors of oil spills too. During the Soviet era, when there was a rush for oil, few cared about the environment. At the Arctic Frontiers conference in Tromsø in 2008, I heard the deputy governor of Khanty-Mansiysk, Gennady Oleynik, talk about the old days. The name "Khanty-Mansiysk" doesn't mean a lot to Westerners, but it should. Travel 2,000 miles east from Moscow across the Ural Mountains and you'll be there, in the vast West Siberian Plain crossed by the meandering Ob River. Bigger than California but with a population of just 1.5 million, Khanty-Mansiysk produces 60 percent of Russia's oil; that is more than any nation except Saudi Arabia.

"It is an open secret that as a result of the human activities in the north we often see irreparable damage inflicted on nature and the environment," said Oleynik. "We have to admit that in the beginning we really did not give any attention to the issue of environmental protection, we felt nature was powerful enough to endure any impact and the livelihoods of the indigenous people would not be under threat from the oil industry." That has changed, he said. "There is a Russian saying 'do not dig a hole for others to fall into.' Resource management in the north should not become like digging a hole into which mankind shall fall."

Some of that past damage was catastrophic. In the neighboring Komi Republic, a thirty-two-mile section of pipeline fell apart in 1994, gushing more than 100,000 tonnes of oil. After a dam failed to contain the spill, over a square mile of land was contaminated and oil rushed into the Kolva and Usa rivers. The fish have yet to fully recover. North of Komi, in the Nenets district that borders the Arctic seas, a gas well blew out in 1980 and continued to burn 2 million cubic meters of gas a day for six and a half years. In a spectacular piece of Soviet hubris, a tunnel was dug close to the blowout and an atomic bomb exploded underground to extinguish the burning gas. It failed, but added a crater and more polluted land.

How do you know that spills are not happening now, a Westerner in the audience asked the deputy governor? "Satellite monitoring systems allow us to discover any spill instantaneously," he replied. "Recently there have been no spills on the territory of the Russian Federation that can be compared with the spills in Alaska, the huge spill from BP's corroded pipeline." (BP spilled 260,000 gallons of oil onto the tundra in 2006.)

Russians don't much like to be lectured by Westerners. Russia now pumps about as much oil as Saudi Arabia, otherwise the world oil leader, and more gas than any other nation. In 2007, the energy sector accounted for one-third of Russia's GDP, 60 percent of its exports, and half of all government revenue. Russia's wealth and power come from energy. With existing oil and gas developments passing their peak, Russia must develop more. The Arctic seas are where Russia is going, lawsuits do not stand in its way, and warnings of risks from environmental groups are not slowing it up.[5]

My introduction to Russia's offshore plans came in neighboring Finland, which is totally dependent on Russia for gas supplies. Aker Arctic, a Finnish company that has designed more than half the world's icebreakers, invited me to Helsinki to their annual "Arctic Passion" seminar. Aker management joked that Russia was responsible for their success. After the Second World War, Finland lost territory to the Soviet Union and access to its only ice-free port. So the Finns had no choice but to become expert at designing icebreakers to get through the frozen winter Baltic Sea.

Aker Arctic has the world's most advanced ice-model testing tank: essentially a giant swimming pool, half as long again as an Olympic pool, with glass panels in its sides and bottom; all sitting in a huge refrigerated building. In the pool you can grow a layer of floating ice of any thickness, complete with pressure ridges if you like, and then drive a model ship through it to see how it copes. It was a wonder to be able to sit below the tank and watch a ship plow through the ice above me. I quickly began to appreciate the subtleties of icebreaker design. It's not enough just to build a strong bow to break through the ice; the hull shape, propeller position, and the ship's wake are all tuned to make the broken ice flow away from the ship so that it can keep moving smoothly.

Ice men from all over the world had gathered at Aker Arctic. And I mean men. When Finland's minister of foreign trade, Paavo Vayrynen, opened the seminar, he began "Ladies and Gentleman" and then looked around the seventy or so people in the room and corrected himself to "Gentlemen." The world of Arctic icebreaking is a world of men and quite a few of the icebreaker captains looked as if they could easily wrestle a polar bear to the ground.

At that seminar I learned that just as in Alaska, a key issue in the Russian Arctic is how to get the oil and gas down south to where they are needed. The choice is ships or pipes. Alaska tried out a giant icebreaking oil tanker back in 1969. The *Manhattan* sailed to Alaska and back through the Northwest Passage, sparking that famous spat with Canada that has never been resolved, but tankers were abandoned as too costly and the trans-Alaska pipeline was built instead.

Thanks to those oil riches in the Khanty-Mansiysk District, Russia already has an enormous pipeline system farther south. Oil can leave Khanty-Mansiysk through a trunk pipeline and travel west to the port of

Primorsk, just outside St. Petersburg on the Baltic Sea. From here it's a short tanker ride to the refineries of Europe. That pipeline also goes east and will one day join the East Siberian–Pacific Ocean (ESPO) pipeline, which is being built west from Russia's Pacific port of Vladivostok. Russia will then have the power to decide whether to export its oil to western or eastern nations. That pipeline system is now pushing branches north to the two regions bordering the Arctic Seas—the Nenets District and its eastern neighbor, the Yamal Nenets District—where oil and gas fields are already busy on land.

But pipelines alone are not enough. I learned in Helsinki that oil from onshore fields close to the sea was being shipped out of the Arctic aboard new designs of icebreaking oil tankers, supported by new icebreakers. More ships are being built and will carry oil from new offshore oil fields in the near future. A little later, gas will be piped ashore from far out in the Arctic seas and travel onward by pipeline to Europe and anywhere in the world aboard liquefied natural gas (LNG) carriers. A new Arctic oil industry is being born, with Russia driving it forward and many foreign partners adding expertise and investment.

I learned that three big Russian projects are changing the Arctic: Varandey, Prirazlomnoye, and the truly ambitious one, Shtokman.

Varandey is just an oil-loading terminal, you might say, so why get

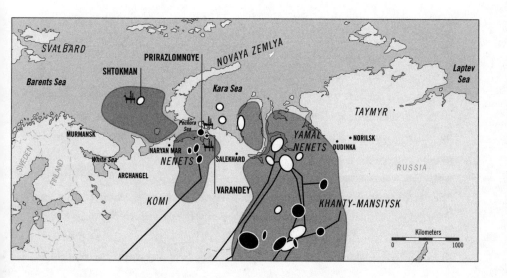

excited about it? In a joint $4 billion project with Conoco-Philips, oil from the fields in Nenets Province flows out through an undersea pipe to a terminal fourteen miles offshore at Varandey. The terminal is a giant conical platform, piled to the seafloor, with a turntable and hose on top. Tankers can pull up here, fill up with oil and head off. In winter a couple of new icebreakers, built in Singapore, keep ice away from the terminal so tankers can maneuver easily. What's new is that the tankers have been built to travel through ice on their own, without icebreakers to help them, and can run a continuous shuttle service, summer or winter, to the ice-free port of Murmansk from which the oil can travel onward. Varandey plus its tankers is a model for the export of Arctic oil in frozen seas.

The icebreaking oil tankers aren't like normal ships. They have two bows of different shape, rather than a bow and a stern. The ship can go forward through thin ice using a more conventional bow at the front of the ship, and backward through really thick ice, using an icebreaker bow at the back of the ship.[6]

Reko-Antti Suojanen, one of Aker Arctic's naval architects, explained to me that the idea may sound strange but it has a history. He pointed to a few of the ship models that sit in glass cases all around their building. Three of them (the *Tarmo*, *Karhu*, and *Tor*) were chunky little red-and-black icebreakers; all had that overhanging bow, rising from beneath the water in an arching curve. The strengthened bow of the ship rides over the ice and helps break down through it. All three ships also had something strange: a pair of propellers at the front of the vessel, set back where the line of the bow vanished underwater, as well as at the stern of the ship where you'd expect to find them.

Back in the late 1950s when these ships were conceived, designers had recognized that propellers at the front of a ship helped move through ice by pulling the breaking ice down and on past the ship. Two sets of propellers proved an expensive solution, explained Suojanen. A neater design would be a ship that could just as easily go backward or forward with only one set of propellers at the stern. That would make it possible to put an icebreaking "bow" at the stern over the propellers to drive the ship backward in thick ice. That design became a reality with propellers that are driven by giant electric motors and sit in pods under the stern of the ship.

The pods can be swiveled to drive the ship in any direction. When they are moving backward through ice, the flow coming off the propeller accelerates the broken ice pieces around the stern and lubricates the hull, making travel easier. If you hit stronger, ridged ice, the propellers are strong enough to even mill through it.

The new ships can keep going through ice five and a half feet thick. Amazingly, each of the three $150 million tankers went from keel-laying to launch in just nine weeks at a Korean shipyard. In June 2008, Varandey opened and the tankers began shuttling Arctic oil over to Murmansk.

Forty miles offshore, much farther than anyone has gone in Alaska, is the second of Russia's giant Arctic developments. Here in the Pechora Sea at Prirazlomnoye is a huge oil field. The sea freezes over in winter, but even this far offshore the water is only sixty-five feet deep. So a gigantic platform, over 330 feet wide and weighing 100,000 tons, is being built at an estimated cost of over $2 billion at the Sevmash shipyard in Archangel. Construction of the platform is running very late, so the field is unlikely to produce oil before 2013. Once it has been floated into place, it will be set on the bottom and another 400,000 tonnes of ballast dredged from the seabed to hold it down. The platform will supply more of those double-bowed icebreaking oil tankers, this time built in Russia's own yards.

Much farther out, almost 400 miles from shore is Russia's grand prize, the Shtokman field, the second-largest gas field in the world. The Russian energy giant Gazprom, Norway's StatoilHydro, and France's Total are working to bring it into production by the very ambitious target of 2013. Here the water is much deeper (1,000 feet), so the drilling platforms must float rather than sit on the sea bottom; temperatures are brutal; there can be heavy fast-moving ice; huge icebergs sometimes wander by; there are savage hurricane-like storms caused by passage of the much-feared polar lows; it is dark half the year; spray can quickly ice up ships and structures and topple them; the distance to land is at the very limit of a helicopter flight so supply and rescue are serious worries; the pipeline carrying the gas to shore is longer than any built in this environment . . . the list of difficulties is great. We'll look more at the technology in the next chapter, but one thing is not in doubt: this is the hottest groundbreaking project in the entire Arctic, and Russia is driving it forward.

I had wanted to find out where the oil industry would make its first move into the Arctic, and I had a big part of the answer, but there are still two other players to consider. In Canada, the Beaufort Sea and the Mackenzie Delta are back at the frontier after a thirty-year lull. There was lots of exploration offshore here in the 1970s when energy prices were high, and even farther north among the more remote Arctic Islands. On land, enormous quantities of natural gas were found in the Mackenzie Delta and along the Mackenzie Valley. Now the big oil companies have returned. In June 2008, BP paid $1.2 billion for a single five-year exploration license in the Beaufort Sea, next to another parcel bought by Imperial Oil and ExxonMobil for $585 million the previous year.[7]

To reach the natural gas, a 750-mile pipeline running the length of the Mackenzie Valley has been planned for almost forty years without an inch of it being built.

I asked Benoît Beauchamp, a geologist and the executive director of the Arctic Institute of North America in Calgary, what the prospects were. "The oil is not Saudi Arabia and I don't think it's on a par with Alaska's North Slope or with the Siberian shelf onshore and offshore, but it's enough to entice industry to go back up there," he explained. "If the gas pipe is built, a lot of the smaller gas discoveries will all of a sudden become within the realm of possibilities. That pipeline will very much become the economic bridge between the north and south. If you get more exploration, then you develop infrastructure, roads, and attract small businesses to support that industry."

But the pipe may never be built. When plans were considered, in 1977, they failed at a famous public inquiry that generated 100,000 pages of evidence. Justice Thomas Berger, who led the inquiry, concluded that the social consequences of the pipeline would be devastating for the native population whose traditional lands it would cross; it would damage wildlife and create a corridor of development that would create few jobs for the people who lived there. He recommended a moratorium while the land claims, and forms of self-government, were settled for the First Nations of the valley. This time around, with their land claims settled, most of the First Nations have decided to do business. All except one have joined together in the Aboriginal Pipeline Group, which intends to buy a share of the pipeline and then profit from the gas passing through it.

That change may not be enough. Costs have soared, an ongoing inquiry into the revived project is already running two years late, environmental groups oppose it, and if the Alaskan gas pipeline goes ahead, it may kill the Mackenzie project. Two big Arctic gas pipelines may push the price of gas too low for them both to be profitable.

The last ice-bound Arctic frontiers lie in the seas to the west and east of Greenland and off the coast of Labrador. Greenland might turn out to be the big surprise. It has no oil or gas development, no industry, and there has been hardly any exploration. But the USGS picked out its seas as full of potentially good prospects. ExxonMobil and Chevron have bought leases there as have lots of smaller oil companies, including Canada's Husky Energy which has built a reputation with successful developments in Newfoundland's seas where there are plenty of icebergs. And Greenland wants to see oil development and the cash it can bring to its budget.

The list may sound long, but it's still only a start. Farther in the future, in areas of challenging remoteness, are the giant gas fields of Canada's Sverdrup Basin and three large areas off the untouched Siberian coast in the Kara Sea, the Laptev Sea, and the East Siberian Sea. Even in these areas, Russia's Ministry of Energy has ambitious plans and has already commissioned concept designs from western companies.

When I left Alaska, I was wondering whether any oil company would successfully move into the Arctic seas. After looking at a wider world the answer is that it is already happening. The big question that underlies the conflict in Alaska is whether it is possible to have both an offshore oil industry and a healthy Arctic environment. If I had to bet on where the answer to that question might come from, it would be Norway, a nation that has not been mentioned so far. Its dependence on oil and gas (half of exports and almost a third of government revenue) approaches that of its neighbor, Russia, but it is accompanied by a public passion for protection of the environment and enormously rich fisheries in clean waters. Norway has to find solutions, and its sophisticated oil industry is determined to move into those Arctic waters. The Snøhvit gas project, well inside the Arctic Circle in the very north of Norway, has already been completed, and all its wells have been hidden away at the bottom of the sea in a technological tour de force. Next up is the Goliat oil field, being developed in the far north by Eni of Italy, and very closely watched by environmental groups.

There's potential for endless conflict and there are plenty of savage attacks on the performance of StatoilHydro, Norway's big oil company, inflamed by spills in the Barents Sea in 2008. But Norway is a small place and consensus is easier to reach without protracted legal action. In 2006 its government decided to try to square the circle of oil development and environmental protection with a holistic, ecosystem-based management plan for the Barents Sea and the adjoining areas off the Lofoten Islands.

The goal for the oil industry is "zero discharge," that is, nothing whatsoever can be spilled into the sea from drilling rigs—no waste, no drill fluids, nothing. The fishing industry is being more closely regulated. Vulnerable areas, and anywhere with ice, are specially protected with oil developments kept out altogether for now. Scores of indicators of environmental health, ever evolving in their sophistication, are monitored and brought together to provide regular overall pictures of the state of the ecosystem. "What is unique about the Norwegian example is that it is an attempt not just to do ecosystem-based fisheries management, but it is trying to integrate all activities. It is a matter of necessity in Norway because our waters have multiple uses," explains Erik Olsen, a senior scientist at the Institute of Marine Research in Bergen, which provides advice to the government. A similar plan for the much larger Norwegian Sea was added in May 2009.

The plan will be revised every four years. In the 2010 revision, the oil industry is pressing for the opening of waters around the Lofoten Islands, a particularly rich area for oil and the spawning ground for many commercial fish. That is very controversial. "Very few systems are totally fail-safe. When the oil industry wants to operate in the most valuable areas we have, where the wealth of the ecosystem is produced, it raises great concerns with me as an advisor to the government, because a large spill could have dramatic long-term consequences. The worst case scenario could set back populations of cod or herring tens of years and close the fishery for a long time," says Olsen.

Holistic management plans don't make issues go away. There is not and never will be a wholly risk-free way to take offshore oil from the Arctic. The arguments are about risks and benefits. Their assessment ultimately falls into the sphere of human values and the same values will never be shared by all. But a holistic management plan can place the issues very clearly before the public gaze and provide objective assessment of what is happening. In that, Norway has been successful.

Thanks to a separate program, MAREANO, even the sea bottom is coming on public display in Norway. Cameras are being towed around the Barents Sea to make a video record of huge areas of the sea bottom and its inhabitants. "We need better maps and mapping of the bottom fauna to be able to pinpoint the most valuable habitats so the project is very important to the Barents Sea management plan," says Olsen. Human impact is made very visible. In some areas, the bottom is everywhere criss-crossed by extraordinary numbers of trawl marks from fishing boats. One trawl mark per eighty feet is common and in the worst spots there is a trawl mark on the seafloor every thirty feet. Other areas are untouched and alive with sponges, starfish, sea pens, sea fans, and anemones. There is a vast cold-water coral reef as rich as any tropical rainforest, the true extent of which was discovered by oil industry exploration submersibles. Touring the project's interactive maps of the sea bottom and stopping here and there to pull up pictures and videos of its luxuriant life—and the hidden damage that has been done to it—it becomes easy to see how intensive monitoring can be made to work. At the same time, the stunning beauty of the seafloor life makes you realize why it is so important that a balance must be achieved, and why it is so hard to argue with those who believe that some areas will really have to be protected forever.

Chapter Thirteen

HOW FAR CAN OIL GO?

In the summer of 2004, three of the world's greatest icebreakers set off toward the North Pole. In the lead was the Russian *Sovetskiy Soyuz*, then the largest icebreaker in the world. With its 75,000-horsepower nuclear engine, the 500-foot-long *Sovetskiy Soyuz* can drive steadily through a field of ice that is a little over eight feet thick. In second place was the Swedish icebreaker *Oden*, diesel powered, 400 feet long, and able to break six and a half feet of ice. Following behind came the ship that they were there to protect, the *Vidar Viking*, 260 feet long, with a four-foot icebreaking capability that in lesser company would make it a powerful vessel.

The fleet passed the North Pole and went on to the Lomonosov Ridge, the chain of underwater mountains that Russia, Canada, and Denmark all claim as an extension of their territorial waters. Settling political disputes was not on the agenda for this scientific expedition. The fleet was here to drill into the ridge to gather data on the past climate of the Arctic that might help shed light on its future.

At the Lomonosov, the ice coverage was nearly complete, $^9/_{10}$ to $^{10}/_{10}$, a mixture of thinner first-year ice and floes of heavy multiyear ice. The *Vidar Viking* dropped a drill string though a moon pool installed in its hull for the project. Now it was essential that it remain exactly where it was, never drifting more than 150 feet off station, or the drill string would snap. The *Sovetskiy Soyuz* moved into position for "ice management," as it is known in the business, where it could intercept the advancing ice fields. There it circled and looped endlessly, smashing the incoming ice into smaller floes. The *Oden* provided close support, breaking up and pushing away the chunks of ice as they closed in on the *Viking*. Further upstream, helicopters dropped buoys equipped with global positioning devices onto the ice to transmit advance warning of the direction from which the ice would attack. Hundreds of radar reflectors were placed out on the ice, too, so that satellites flying overhead could help to track the ice.

Just one week after arriving, the coring team drilled down 300 feet, a tremendous achievement given that no one before had drilled more than the top few feet in the central Arctic.[1] The ship's log reads: "Tough struggles against heavy ice . . . *Viking* almost lost its position and risked breaking the string but intense efforts by the two assisting icebreakers and a change of ice drift saved the situation. . . ." Two weeks later, the fleet had to depart because winter was coming. They had set many new world records, including drilling to over 1,300 feet and managing to keep the drill ship "on station" in heavy ice for 125 hours. They had the cores they wanted. Their results were to help rewrite the early history of the Arctic, showing that 50 million years ago the Arctic was warm and capped by freshwater.[2] A floating fern called Azolla grew here in such massive blooms that it may have triggered a cooling of the entire planet by sucking carbon dioxide out of the air and weakening the greenhouse effect. That Azolla died long ago and was likely transformed into oil and gas deep below the bottom of the sea.

The teams of scientists on this trip were not much interested in oil, but the oilmen certainly knew all about their expedition. And with good reason: in asking how far the quest for oil can go into the deep Arctic and what technology is available, this expedition is a good place to see the outer limits.

I've been to quite a few Arctic oil conferences and at some stage somebody will show a picture of the three icebreakers at work, taken from a plane flying high overhead. The picture shows a white, frozen sea fractured into hundreds of ice floes amid rubble ice. You can just pick out the *Vidar Viking*, hanging onto its station in the ice, by the little patch of open water at its stern where the ice rubble has passed. The magnificent *Sovetskiy Soyuz* and the *Oden* are easier to spot; their bigger wakes each make a protective curve of open water in front of the *Viking* while the ice field comes on against them.

When this picture comes up, the room always falls silent. These are the most powerful icebreakers in the world, but their efforts look puny amid the Arctic ice. Imagine drilling not a thousand feet to pull out a scientific core but tens of thousands of feet beneath the ocean floor to reach oil imagine staying on station not for a snatched few weeks before winter returns but for years on end to pump that oil to . . . what, exactly? A pipeline? Not at these distances and depths. To storage on board a ship (a "floating production storage vessel," or "FPS" as it's known in the trade) so that tankers can shuttle back and forth to pick up the oil? State-of-the-art icebreaking oil tankers can't come close to what *Sovetskiy Soyuz* and *Oden* fought in the height of summer. Wait until global warming melts the ice away? Summer in the central Arctic lasts barely a month and then the ice and perpetual dark return.

Whenever I chat with people at these conferences they all agree that stories of a "cold rush" for the riches of the central Arctic lie very far in the future. Right now, they want to look at the technology that will let them put a toe in Arctic waters, in the most accessible areas, as close to land as possible, preferably in shallow water with little heavy ice. That feels comfortable. There are ideas for working farther out, but the combination of deep water and multiyear ice far from shore produces long, reflective silences. A structure will have to withstand both ice and big waves, dodge icebergs, and somehow get its oil or gas to shore: the reaction then is always, "maybe in twenty years, maybe a lot longer, and only if the price of oil is very high."

The U.S. government's Minerals Management Service (MMS) shares the oilmen's view of the hard parts of the Arctic. In 2008, it published a 339-page overview of Arctic oil technology, focusing on Alaskan waters but reviewing developments around the circumpolar world.[3] "Floating production systems for the Beaufort Sea, Chukchi Sea, and North Bering Sea are not considered to be technically feasible, even with continuous ice management," it concluded. "No floating production structures could be economically designed to stay on station with multiyear ice loads found in the Beaufort and Chukchi seas."

The deep Arctic has not been conquered yet, and talk of imminent fights over the riches at the North Pole is nonsense. To explore what has already been done, what might be possible soon, and what might have to wait for decades, I turned to the principal author of that MMS report, Michael Paulin, president of IMV Projects Atlantic (IMVPA) in St. John's, Newfoundland. IMVPA is part of the Wood Group, a multinational engineering group with headquarters in Houston but built up by a couple of fishermen from Aberdeen in Scotland who seized the new opportunities emerging in the British North Sea oil industry thirty years ago.

The technology for the coming "second oil age" builds on the technology of the oil age which began in the 1960s with the discovery of oil on Alaska's North Slope and boomed in the 1970s when the Yom Kippur War sent oil prices soaring. After oil prices fell to $20 a barrel in the late 1980s and refused to go up again, Arctic offshore oil exploration ground almost to a halt.

The first time around, oil companies began venturing offshore into the Beaufort Sea "by working with Mother Nature," as Paulin put it, building islands made out of ice. Nothing could be simpler in principle. Just spray a lane of seawater onto the shore-bound fast ice, let it freeze and harden, and keep going until you have built an ice road out over the sea ice. At the end of the road repeat the process until you have created your very own island to support a drill rig. You just need to be careful that the island is firmly grounded and will not unexpectedly turn into an ice ship. More than twenty such ice islands were built all over the Beaufort Sea during that first oil age. Of course, the islands melt away in summer so they are suitable only for an exploratory drilling campaign. What if you want something more permanent?

"One of the simple things, well relatively simple, because none of it is that simple, is to build a gravel island," explains Paulin. "In the shallow Alaskan Beaufort Sea you have North Star, Ooguruk, and another one coming called Nikaitchuq. North Star is built in the deepest waters—it is about fourteen meters [forty-six feet]." To build North Star, six miles from land, a winter ice road was put in, then a convoy of trucks ran along it and dumped over 100,000 loads of locally quarried gravel into the sea at its end.

Between 1976 and 1990, when the first Arctic oil age was going flat out, seventeen gravel islands were built on the Alaskan side of the Beaufort and more than thirty on the Canadian side, some by simply dredging the sea bottom and piling up mud, rather than trucking gravel out from land in winter. But gravel islands have their limitations. They settle into a giant pile with long sloping sides, so in deeper water staggering volumes of gravel are needed. "Gravel becomes uneconomical beyond a water depth of about twenty-five to thirty meters [eighty to one hundred feet]," Paulin explains. "That is when you can start to look at ground-bottomed structures."

Ground-bottomed structures are built of steel or concrete and are essentially artificial islands with vertical sides. They are towed to location, sunk on the sea bottom and filled up with dredged ballast to keep them in place. What guides the design? "In the Arctic they have to be installed in the open water season, during summer. The season may be very short so it's important that you can just bring it in and set it down on the bottom fairly quickly. Even when there is no ice today, you can have ice tomorrow, suddenly blowing in from the polar pack," warns Paulin. "Your foundation—the geotechnical condition of the sea bottom—is very important. The structure must not slide or tip over. Obviously water depth is very important," he continues. "Then there are the sea-ice conditions: is it first-year ice? Multiyear ice? A mixture? In places like Sakhalin in Russia, you only get first-year ice which is not as strong as something that has trailed around the Arctic for two, three, or more years. Do you get any icebergs? Here on the east coast, we can get a lot of icebergs. There are not so many in the Beaufort Sea, but you can get enormous ice islands and icebergs that trickle through from the High North. These are huge pieces of ice, weighing maybe a million tons. How do you design for it? Then you need to know what sort of waves you have. What does the structure need to carry on top of it? Is there some kind of storage requirement? If you are looking at a platform with no pipeline to shore, you would have to have some storage for oil. A lot of that will dictate what your structure looks like."

The Molikpaq was the giant that took the ground-bottomed idea to its limit. Built in 1984, it was a colossal steel octagon, 300 feet across, 26 feet deep, and it weighed 43,000 tonnes even before it was sunk to the bottom and filled with sand. The "Giant Wave," as its name translates from Inuktuk, did good service until 1990 when the era of low oil prices arrived. Almost a decade later it was brought back to life and towed away to Russia's far eastern Sakhalin field. En route, it stopped to be "mated" with a second octagonal structure so that it then stood some 50 feet taller and 15,000 tonnes heavier. Now it sits ten miles out from the Russian coast in a hundred feet of water, collecting oil from the wells beneath it. When the winter ice melts, tankers shuttle back and forth, offloading stored oil. Some 120 people live on board.

I've seen a film of the new-look Molikpaq taken from an icebreaker circling it. From a distance, the sheer reach of its smooth steel walls and the cluster of towers, derricks, and enormous storage buildings on its topsides make it look far more like a medieval castle than anything from our century. The engineers who build these structures are perhaps the true heirs to the castle builders: they rely on size, strength, and mass to subdue the ice and storms, and they impress by the gigantic scale of their creations.

An even bigger platform will be coming soon. That's the giant structure being slowly constructed at the Sevmash yard in Archangel, and intended for Russia's Prirazlomnoye field, forty miles offshore in the Pechora Sea. This monster swallowed another giant platform, the Hutton, which was no longer needed in the North Sea and towed off to Russia. Although the Hutton weighed 37,000 tonnes, it is now just a part of a bigger structure that is 400 feet across, weighs 100,000 tonnes, rises 400 feet above the sea and will carry a crew of 200 workers. Once full of ballast it will weigh 500,000 tonnes. Its only rival is the Hibernia platform, the largest bottom-grounded structure in the world, which sits off Newfoundland and weighs over a million tonnes.

But even a ground-bottomed structure has limits, Paulin explains. "As the water gets deeper, the tower gets taller. Farther out from land, there may be harsher, stronger multiyear ice bearing down on you. A fixed structure is only so good, so far. Then you have to have a full subsea tieback to shore or a floating structure."

Designs cannot simply be borrowed from deep-water fields like the Gulf of Mexico, where thousands of drill rigs operate safely in hurricanes.

These rigs are often built up on "jackets," latticeworks of welded steel pipes rising from the sea bottom. Some can be enormous: the *Bullwinkle*, launched in 1988, is 1,365 feet tall. Such open structures can handle waves well enough, but not the incredible load of a moving field of thick ice.

So what is a "full subsea tieback"? We pass from the ages of castles to that of high technology. Subsea tiebacks are the pipelines that directly connect a well on the seabed (perhaps hidden in a "glory hole" to stop anything running into it) back to the shore. Paulin points me to Ormen Lange and Snøhvit (Snow White) as the two projects, both gas rather than oil, that have pushed the limits of subsea tiebacks.

Ormen Lange and Snøhvit, which came on stream in 2008, are the pride of Norway's StatoilHydro. The energy giant reckons that they set the agenda for low-environmental impact work in northern waters. Ormen Lange carries gas from a set of wells in the North Sea nearly seventy miles to land, while Snøhvit up in Norway's Arctic seas off Hammerfest carries gas ninety miles. Two of StatoilHydro's senior engineers, Jan Inge Dalane and Roald Byhre Sirevaag, talked over the basics with me.

Down beneath the water is a big steel frame (around seventy to a hundred feet across) that is fixed to the seabed and connected to the wells, with sloping edges to stop trawlers snagging their gear. To operate the system, you need a pipe to carry the gas away and others to take electric power, control signals and anything you might need to add to the flow (to stop water in the pipe freezing, for example). Subsea wells aren't unusual—StatoilHydro alone has almost 500 around the world—but such long tiebacks to shore are special.

One of StatoilHydro's goals is to make even longer tiebacks possible. The biggest problem was a surprise to me. "We need to develop power transmission systems," explains Sirevaag. "You can use AC power up to some hundred kilometers, but you can't use it for longer distance. You need to go to a DC system and quite a significant development effort is needed to make that happen." Pumps or compressors placed on the seabed to push oil or gas to the shore will need lots of power. In the future, it may be possible to carry oil over a hundred miles and gas two or three hundred miles like this. That would enable oil and gas to be taken away from under the ice, even in deep water, provided drill ships could break through the ice for long enough to install and maintain the equipment. Thick, permanent ice is a different matter.

"What happens if you are under ice for nine months of the year?" asks Paulin. "And what do you do to work over your wells or correct something or repair something? That's a challenge. In the Gulf of Mexico and other places where they do very deep water subsea installation, if you want to put a clamp on a pipeline or cut out a section of pipeline and weld it in, those things all can be done using remote vehicles. If you are covered with ice, how are you going to do that? You'd better think about it because you need to prove that you could do that in the Arctic."

The alternative is to build a floating structure that can face icebergs and ice, rather than hide from them under the sea. No one has done that in the Arctic yet, but one answer has taken shape not so far from Paulin's office. Out on Newfoundland's Grand Banks is the White Rose field where two floating oil platforms connect to wells 400 feet deep in the waters beneath them. They don't face heavy ice, but they do have to deal with giant icebergs that come wandering down from Greenland, the kind of iceberg that sank the *Titanic* just 375 miles from the shore here. The platforms have been designed so that when a big iceberg approaches, they can shut down the flow of oil, disconnect a flexible riser coming up from the well on the sea bottom, and move to safety. Since starting work in 2005 they haven't had to flee from an iceberg once, though they were on high alert in 2008. That's partly because there is an earlier line of defense. Powerful tugs are on standby to go out and lasso incoming icebergs by circling them with a giant rope. Although icebergs easily weigh over a million tons, a slow steady pull over a few miles of drift can put them on a safe course. But this is not an option where icebergs mingle with heavy ice in the High Arctic.

That's not the frontier that oil companies are planning to tackle right now. "Once you get out beyond the transition zone, beyond the fast shore ice and into moving ice, even in thirty to thirty-five meters [one hundred to one hundred and fifteen feet] of water depth, that is the next frontier," says Paulin. "Deeper water with heavy ice is the far frontier where there are no off-the-shelf solutions, just concepts."

There is just one project on the Russian side of the Arctic that is trying to make a leap.

"Shtokman" is a name that excites everyone in the Arctic oil and gas business. Russia's Gazprom, Norway's StatoilHydro, and France's Total have come together to try to exploit a gas field, 370 miles off the Arctic coast, much farther out than anyone has gone before. The Shtokman project is a "step up" from what has been done before, says StatoilHydro's Dalane. "On the Grand Banks the main issue is icebergs, but in Shtokman we also have severe sea ice, so the floating facility will operate in relatively harsh conditions." "Relatively harsh" is an understatement. The water is quite deep, at 1,100 feet. The area can be covered in ice, and some of it is old, strong ice, ten feet thick with ridges. But there is open water too and big waves, up to sixty feet high. In 2003, fifteen icebergs were recorded, two of them weighing more than 3 million tons. Then there is the intense cold, with winter temperatures of -35°C so sea spray can quickly freeze to superstructures and weight them down with ice. All work must be done in totally enclosed areas, and that increases the risks of gas explosions. In winter it will always be dark, placing a great strain on crew, especially as the field is at the very limit of a helicopter rescue flight. And there are the polar lows, the unpredictable and intense storms that can spring up in the Arctic that could cause big problems for the support vessels. The only plus is that it is a gas project; pumping gas to shore is a lot easier than pumping oil, and the environmental risks are much lower.

The technical challenges are all worth taking because this is not just any gas field but the second-largest gas field in the world, capable of producing gas for Europe for half a century. The solution is a ship-shaped floating platform which can turn to face the oncoming ice, which will be broken up by accompanying icebreakers. If a giant iceberg or unmanageable ice threatens, the ship can disconnect from the field below and move away.

Designing that ship is not easy. Per Kjaernes, who is managing the project for StatoilHydro, described the hard part as "the platform and everything around it" at a conference held by icebreaker designer Aker Arctic in Helsinki. To reinforce the Norwegian view of "hard," he ended his talk with a picture of Roald Amundsen and his men standing at the South Pole after their epic journey in 1911. How quickly would the ship have to disconnect its mooring lines and a gas riser if an iceberg threatened, he was asked. "We need to disconnect in fifteen to thirty minutes," he says. "It's a challenge."

Next comes the problem of getting the gas to shore. "A piece of cake" is how Kjaernes describes the job of laying a 370-mile pipeline across the bottom of the Barents Sea. Still, it didn't sound so easy to me, so I visited J P Kenny, the pipeline company in Aberdeen that had been awarded the initial Shtokman design work. Jonathan Lindsay, the company's international business manager, answered my naive questions. He trained as a mining engineer and "qualified just as all the mines were closing down," he explained, and moved into the oil business.

The only gas pipes that I had ever seen were rigid lengths of steel tube being lowered into a trench on land by a crane. I was wondering how you could bolt all those pipes together at the bottom of the sea. "You are going to be surprised," said Lindsay. "The pipes are made of exotic carbon steel. You can take a rigid piece of pipe and bend it right round a giant reel." He showed me pictures of pipe-laying vessels at work. Pipe comes off the reel through a straightener and runs off the back of the ship in a long curve right down to the seabed, almost as though it were rope. Steel pipe that can be rolled up and unwound seemed miraculous to me. "Up to sixteen-inch-diameter pipe can be reeled but not much bigger," explained Lindsay. "The small diameter reeled pipe is just for short distances. Bigger diameter pipes are usually put down from large lay barges, moved around by tugs."

For these bigger lines, lengths of pipe are welded onto the end of the line on the barge and led off its back, down a "stinger"—a device that supports the pipe in a gentle curve as it descends on rollers. Once again the pipe has to support its own weight as it gradually settles on the sea bottom. "It's a slow process," said Lindsay, "as the ship lays pipe, it inches forward." Another supply barge keeps craning fresh pipe aboard. It won't be any old pipe. Design companies like JP Kenny have to find the right pipe for the fluids it will carry, under the range of temperatures and pressures it will experience and without corroding for a quarter of a century or more.

In the Arctic the ice and the weather mean the construction season is very short, so laying the Shtokman pipe is a challenge. On the seabed, however, there are a whole lot of extra problems. First there are icebergs with deep keels that run into the sea bottom and dig out great gouges. Submarine pictures of some parts of the Arctic show the seafloor crisscrossed with great trenches cut by monster bergs. "At the moment there isn't a design that will allow you to lay a pipe that can be hit by an iceberg," said Lindsay. "The iceberg will wipe it out." An obvious solution is to bury the pipe.

But pipe burying turns out to require yet another set of specialist ships. Some use giant subsea plows. "You have to lift the pipe up and plow underneath and let the pipe go down again," explained Lindsay. "Or you can use water jets. That is a gentler way of doing the job: if the bottom is made of the right material, you can fluidize the trench and pump the stuff out. Then the pipe just settles down in."

That doesn't quite solve the problem, though. In the Arctic you have to watch out for "strudel scour," holes that form in the ice in spring into which meltwater drains. That flow can turn into a torrent and the whirlpool below can drill deep holes into the seabed, exposing buried pipes. With big icebergs and risks from strudel scour you might have to bury pipes very deep indeed. This is tricky, as Lindsay explained. "It is difficult to bury anything more than a few meters below the surface. With a plow it is difficult to cut a vertically shaped box, especially in sand which collapses in again, so you have to cut at an angle. The angle usually has to be 30–35 degrees." To make a deep furrow, you would need a truly gigantic underwater plow.

The last thing I learned is that it is a big mistake to think of the seafloor as a flat and comfortable plain. Lots of specialist survey work is needed before you can even start. The pipeline from Shtokman has to find routes around hills and across valleys avoiding giant boulders that have been dropped from ancient icebergs and, as land approaches, climb a steep rise. Not quite a "piece of cake."

Shtokman is the one project underway that pushes oil and gas exploration to new limits. If it succeeds in its very ambitious target of delivering gas by 2013, it will have raised the limits of what is possible in the Arctic. There are rumors of other projects that may push the boundaries beyond those shallow shelf seas, including in the Beaufort Sea off the coast of Canada where the water is more than 2,000 feet deep and there is ice, although not as tough as at Shtokman. In this region, one big lease runs out in 2012 so there's pressure to begin exploratory drilling by then.

Long before production vessels or subsea production can get going, other special ships have to conduct the surveys, find the oil or gas, and drill the first wells. Only one drill ship has ever been built specifically to work in the Arctic ice and that's the circular *Kulluk*, launched in 1983 during the first Arctic oil age and brought back to life by Shell for the Alaskan drilling program that ran into so much opposition.

New drill ships are emerging only slowly. Houston-based Frontier Drilling is working with Shell to build two in Singapore. With icebreaker bows and ultraflexible steel to protect the hulls from shattering in extreme cold, they will be able to work among the ice using computers to control thrusters which keep the ship in one place. Another ship, the *Stena Drillmax Ice*, is being built in Korea for Aberdeen-based Stena Drilling at a cost of $1.3 billion.

Even these amazing new ships don't necessarily signal a rush for the Arctic. All are designed to be capable of drilling in ultradeep water—the oil world's other frontier. The first of the new ships will likely spend its first five years on charter in deep water exploration in the Gulf of Mexico, according to Frontier Drilling. The *Stena Drillmax Ice*, which will be finished at the end of 2011, could go off and drill among summer ice in the Arctic, but it would be just as happy tackling water 10,000 feet deep off the coast of Africa.

I am still wondering whether there are final frontiers which no oil company will ever attempt, perhaps the kind of place where those three powerful icebreakers struggled to drill on the Lomonsov ridge. So I ask the two engineers from StatoilHydro. "We are technologists, so we are optimists. It depends on the size of the reservoir," says Sirevaag. "And the price of oil," adds Dalane. What if there is thick, permanent ice? "There are concepts for seabed rigs which you could use to drill underneath the ice," says Sirevaag. "The tie-back distance to any kind of infrastructure in ice-free areas would be the biggest issue to solve. Maybe you could use sub-seabed storage and big submarines to go and get the oil and compressed gas." I ask if they are kidding me. Sirevaag replies, "Somebody got to the moon."

Chapter Fourteen

TOO MANY SHIPS, TOO SOON?

O n my desk is an odd little memento which I was given more than twenty years ago. I shared an office in the National Press Building in Washington, D.C., with a journalist who had gone out to Prince William Sound in Alaska to write about the Exxon Valdez oil spill. When he came back, he handed me a small plastic box and said, "Hundreds of miles of beach are like this." Inside was a pebble encased in black, gooey oil. The pebble had been lying in the cold, clear waters of the sound, home to cormorants and otters and hundreds of other creatures until the Exxon Valdez ran aground and spilled 10 million gallons of Prudhoe Bay crude.

After twenty years the oil in the little box is still sticky and the wildlife of Prince William Sound has still not recovered completely. The pigeon guillemot, a close relative of the black guillemot, shows little signs of rebuilding its population; others like the sea otter, the goldeneye, and the black oystercatcher, along with the clams and mussels at the bottom of the sound, are not yet back to full health.

This anecdote might seem an odd place to begin a chapter about shipping in the Arctic but it is an essential preamble. A boom in the numbers of ships traveling back and forth into the Arctic has already begun. There are tankers traveling round the Arctic coast of Norway, ore carriers crossing Siberian and Canadian waters, drill rigs and icebreakers heading into the Chukchi Sea, and, in the fastest boom of all, tens of thousands of tourists traveling on cruise ships everywhere they can go, from the coasts of Greenland to the North Pole itself. All this is happening without new regulations to prevent and deal with shipping accidents or oil spills.

That is making many people worried. In April 2009, the Arctic Marine

Shipping Assessment was delivered to the Arctic Council's top-level ministerial meeting.[1] Four years in the making, and with an executive summary that is almost 200 pages long, it confirms the boom in shipping and
predicts where further growth is most likely. It stressed concerns raised a
year earlier, when experts on oil spills from all over the world met at the
University of New Hampshire. They too produced a report.[2] It was much
shorter and much scarier.

Imagine the following scenario. A cruise ship carrying 1,400 passengers is traveling along the west coast of Greenland. That's not unusual. I've
seen ships that size sneak into a fjord to give passengers a really close-up
view of a calving glacier. This time the ship hits bottom on its way back
out to sea and begins to sink. The passengers, along with 700 crew, must
abandon ship; most of them are between fifty and eighty years old, some
are infirm, and a few are in wheelchairs. It is mid-September and cold
enough that passengers have a 4.3 hours functional time (the time before
they are incapacitated by cold). Within that functional time there is only a
67 percent chance of survival.

Small fishing vessels and a coast guard cutter arrive within a few
hours to help, but they cannot take on the passengers; other cruise ships
that could do so are a day's journey away. The sinking ship begins to spill
its fuel oil which drifts toward the Ilulissat glacier World Heritage site.
The nearest pollution response vessels are in the United States and would
take two to three days to arrive. Almost nothing can be done to save the
passengers or contain the drifting oil.

That is one of five scenarios which were put together for the workshop.
Reading them all, I wondered if this weren't a convention for Hollywood
disaster movie scriptwriters rather than a scientific meeting. In other scenarios, an icebreaker catches fire and rams the drill ship it is supporting
within the disputed U.S.-Canadian border zone, causing it to lose fuel; a
tug laden with fuel and towing a barge loaded with a cargo of explosives
for the mining industry loses power in the Chukchi Sea near a haul-out
spot for walrus, the tug spews fuel while containers of explosives wash
up on shore; and a tanker smashes into a fishing boat near the disputed
Russian–Norwegian boundary and spills huge quantities of oil while the
fishing boat sinks.

A conference call to the workshop organizers, Amy Merten from the National Oceanic and Atmospheric Administration (NOAA) in Seattle and Nancy Kinner of the University of New Hampshire, confirmed what I feared—this could all happen today. "That's correct," said Merten. "We had a committee that helped us put together the scenarios and the goal was to make them real, something we could consider right now, knowing that we are going to see more and more activities in the Arctic water." Even that cruise ship scenario where hundreds of older people may die? "Just to reinforce," says Kinner, "the group that looked at that cruise ship scenario consisted of people who had familiarity with those cruise ships. That scenario was not really extreme. The statistics about passengers are the reality of the cruise ship industry."

Accidents in polar waters have already happened. In 2007, the cruise ship MS *Explorer*, with 150 passengers and crew, sank off the coat of Antarctica. I'd been in the same area in a ship the same size earlier that year, so I was transfixed by the images of the bright red ship heeling over amid the ice floes and then disappearing. Everyone was rescued thanks to the lucky presence of another cruise ship, the Norwegian vessel *Nordnorge*. "In that situation they had everything on their side," says Merten. "They had pretty calm weather, they had another vessel that was right there that was able to offload the people. The problem is exponentially greater when you're talking about 1,400 people."

Bigger ships have been in trouble in the past. Back in October 1980, the cruise ship *Prinsendam* caught fire and sank in the Gulf of Alaska with 510 passengers and crew on board. Once again, everyone was saved, including many older cruise passengers, thanks to a piece of luck. A supertanker, the *Williamsburgh*, was nearby and provided a landing pad for U.S. Air Force helicopters.[3]

The point of these scenarios was to push for urgent action. If the Arctic is going to see more shipping, it needs "ports-of-refuge" (there are virtually no ports across the top of the North American side of the Arctic; the Soviet Union built a chain of now run-down ports from Murmansk on east through Dikson, Dudinka, and Tiksi to Pevek, the most northerly port in the world). It needs better charts, better weather and ice forecasting, internationally coordinated search and rescue, survival equipment on ships, salvage tugs on standby, oil spill countermeasures in place and equipment ready to go, new international rules on ship safety, and, of course, better knowledge of the impact of spills and technology to deal with them, especially among broken ice.

The AMSA report was able to take a one-year snapshot of traffic in the Arctic, drawn from records of all the circumpolar nations. It recorded the expected mix: ships carrying bulk cargoes to and from mines and oil fields at the edges of the Arctic seas, ships and barges carrying supplies for Arctic communities in the summer, tourist ships making multiple summer voyages as far north as they can go, and many small fishing vessels, especially in the Bering and Barents Seas. Approximately 3,000 vessels were operating in the Arctic that year (the number doubles if ships just cutting through the Aleutian Islands on the Arctic route between the U.S. West Coast and East Asia are included).

But those numbers come from 2004, the latest year for which data could be found, and all the signs are they are already out of date. A boom is just beginning. For the bulk carriers, it is at the western end of the "Northern Sea Route." That route leads from Norway right across the top of Russia and on to the Pacific, but it is on the European end that traffic is building now, as tankers and ore ships travel back and forth to the oil fields and mines along the Siberian coast and around the Norwegian coast to foreign markets.

Vardø, a small town perched on the most northeasterly point of Norway, with a view out to the Barents Sea and across the top of the Kola Peninsula to Murmansk, is the place to see the ships. Norwegians have known for a long time that a traffic buildup was coming and have been deeply worried about the consequences of a spill for the rich fisheries of the Barents Sea. Unlike almost any other place in the Arctic, they have already executed a plan to spot trouble on their coast and have built a new tracking center.

I talked to Ståle Sveinungsen, director of the Vessel Traffic Service. His team sits surrounded by computer screens with radar images of the surrounding seas, has access to the automatic ship-identification systems, and is ready to call the three patrolling tugboats, coast guard, emergency services, and the Royal Norwegian Navy as necessary. "If we have an emergency, we run through a set of procedures, it's like the cockpit in an airplane," Sveinungsen says.

Vardø tracks the movement of any ship larger than 5,000 tonnes [around 300 feet or longer] through Norway's 200-nautical-mile economic zone, as well as ships traveling around the islands of Svalbard, farther to the north. The three tugs ply the zone ready to help and act as the center's "eyes." Out at sea, an agreement with the International Maritime Organization (IMO) keeps traffic in two separate lanes and more than thirty nautical miles off the shore (rather than the usual ten nautical miles). "There will be more and more activity in this Russian area and to get there, you have to come along the Norwegian coast.[4] We are ready for this traffic. The coast is much safer now," says Sveinungsen. The first sign of trouble is when a ship comes to a halt. "We see two or three vessels every month drifting, many of them have just a small problem, they are changing filters or something on the main engine and after a few hours they start sailing again," he says. "We can see them out there and we have time to send a tug, so it's not a big problem anymore."

Around the corner from Vardø, Russia is commissioning innovative ships, including tankers and ore carriers with the "double acting" design that I had seen in Aker Arctic's giant testing tank in Helsinki. These are the ships with two bows, one that is close to the conventional V-shape and suited for more open water, and one that is a spoon-shaped icebreaker bow.

They are expensive to build compared to a conventional ship, but they don't need an icebreaker escort to move through ice up to five feet thick. That saves a lot of money (30 percent of running costs according to Norilsk Nickel, one of the operators) and uncertainty. Russia still has by far the largest icebreaker fleet in the world, including five nuclear-powered icebreakers of tremendous strength.[5] But they are growing old. Only one of its atomic breakers has a planned life expectancy beyond 2020; that is the *Fifty Years of Victory*, the most powerful icebreaker in the world, launched in 2007. Rosneft, the Russian company which looks after the fleet, says it will build more nuclear icebreakers and has a budget for the first of them.

Already in service among the "icebreaker-independent" fleet with the double-acting design are three tankers built in Korea shuttling between the Varandey oil terminal off the Siberian coast and Murmansk. Two more tankers are being built in the Admiralty Yard outside St. Petersburg (the first will launch in 2009) to serve the giant Prirazlomnoye oil platform.[6] And five ore carriers are already running, the last of which arrived in March 2009. Each is quite small (14,500 tonnes deadweight) but the cargo of nickel they carry is very valuable.

The ore carriers are owned by Norilsk Nickel, the world's biggest producer of nickel and palladium, and replace an earlier design of icebreaking ore carriers of the Norilsk class.[7] At one end of their run is the port of Dudinka, linked by train line to the mines of the legendary city of Norilsk, one of the most polluted places on the planet. More than 16,000 prison laborers died working in the mines and building the city before the Soviet gulag system came to an end. The city is now closed to foreigners, but it is still easy to visit: virtually, that is. Residents have posted videos of the city on the Web, taken from automobiles driving around town summer and winter, and they provide fine views of the city's legendary smokestacks and Soviet-era plant, all set to rock music. At the other end of the ore carriers' run—1,343 nautical miles away—are Murmansk and the smelters of the Kola Peninsula.

Coming to the Russian Arctic fleet in the next decade will be liquefied natural gas (LNG) carriers—the odd-looking ships that are built around a line of spherical gas tanks. Russia plans a liquefaction plant near Murmansk to handle the gas from the pipeline to the Shtokman field and another among the reindeer on the Yamal Peninsula (for once the Arctic offers an advantage; the bitter cold makes it much cheaper to liquefy gas up here than in the heat of the Persian Gulf). Double-acting ships may be built that can break through ice and travel on to deliver their cargo to markets in Europe and America, or icebreaking ships may simply shuttle back and forth within the Barents Sea, transferring their cargoes, in port or at sea, to conventional ships that are cheaper to run in open water.

Across the other side of the Arctic, Canada's Arctic shipping specialist, FedNav, has icebreaking ore carriers to serve the mines at Raglan in Northern Quebec and a new ship, the *Umiak*, built in 2004 in Japan for the Voisey's Bay nickel mine in Labrador. The *Umiak* is a conventional ship of great power, capable of ramming and breaking through five feet of ice and carrying 30,000 tonnes of nickel. There will be a shipping boom here only if natural resources are exploited; the enormous Baffinland iron ore project being the biggest among them. And if the Alaskan offshore oil fields take off, the seas there will be busy with seismic survey ships and drill rigs and all the supporting vessels they need, as they will be in Russia as the search for oil moves out east toward the Ob, the Yenisey, and the Lena over the next two decades.

Oil, gas, and mines are clearly making the Arctic seas a busier place, and could make it busier still, but the big question is whether that is just a fraction of the coming Arctic shipping boom. The prospect that sets the imagination alight is a fast route between the Atlantic and the Pacific, especially one through the Northwest Passage, where so many explorers died looking for the fabled route to Cathay.[8] The Northwest Passage is not really one route, as there are five variations that are possible among Canada's many islands, with the easiest to get through being too shallow for big ships.[9] Largely ice-free routes opened briefly in 2007 and 2008, for the first time in history, and a few tourist ships slipped through. It was not until 2008 that the first boat made it through the passage on a commercial voyage. The ice-strengthened 440-foot MV *Camilla Desgagnés* delivered cargo to Inuit communities in September 2008 during a short period when "there was no ice whatsoever."[10]

A tourist route it may become, but the Northwest Passage is the least-favored trans-Arctic shortcut between the Atlantic and the Pacific, with the Northern Sea Route around Russia and the direct transpolar route offering better opportunities. That, at least, is what I'd heard Douglas Bancroft, head of the Canadian Ice Service, say at a shipping meeting back in early 2007. I called him to ask if the summers of 2007 and 2008 had made him change his forecast. "Not at all. In fact, I am even more convinced. We saw areas of open water fairly close to the North Pole last summer [2008].

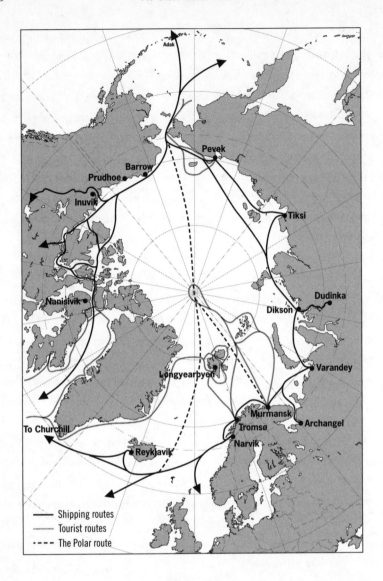

Shipping routes
.... Tourist routes
- - - - The Polar route

The northern sea route we still forecast to open first on a reliable summer basis and then the transpolar route. It is still true now that most of the multiyear ice is packed up to the north of the Arctic archipelago. Because it's becoming much more fractured and mobile, we have the risk of large chunks of multiyear ice, in or near the Northwest Passage. That's an impediment for the shipping." There is nothing worse than chunks of extremely hard, clear old ice, floating at the surface where it is difficult to see, for a ship traveling at speed in seemingly open water or thin ice. A collision may sink the ship. This seems to be what happened to the MS *Explorer*, the tourist ship that sank in the Antarctic, according to the authorities in Liberia where it was registered.[11] The ship was passing through what its crew thought was thin first-year ice, but hit a wall of much harder land ice (from glaciers), holing the hull. The Northwest Passage will be the last place to be free from these dangers, as scattered old ice floes hide in the thinner ice.[12]

So we must return to Russia again. The first shortcut across the Arctic will likely be the old route across the top of Siberia.[13] The route passes from Murmansk over the Barents Sea, through the narrow "Kara Gate" between Novaya Zemlya and the mainland, across the Kara and Laptev seas inshore of Severnaya Zemlya, and then winds among the New Siberian Islands through the Sannikova Strait before passing inshore of Wrangel Island and arriving at the Bering Strait. A nearly ice-free summer passage could have been made along the route in 2007 and 2008, except for the western Laptev Sea.[14] As the ice thins, the route will be able to run farther offshore, cutting out the shallow seas at the Sannikova Strait and eventually heading straight across the ocean on the transpolar route.

On paper, the route looks wonderful. The distance from Europe's busiest port of Rotterdam to Yokohama in Japan is 11,250 nautical miles through the Suez Canal and just 7,350 nautical miles via the Northern Sea Route over the top of Russia; to Vancouver on the west coast of Canada it is 8,920 nautical miles through the Panama canal and 6,980 nautical miles over the top of Russia. When the Soviet Union kept a fleet of icebreakers ready to lead ships through it in convoy, the route was passable all summer—331 ships sailed on 1,306 voyages in 1987.[15] The western end of the route, to Dudinka, has been kept open every year since the winter of 1978–79.[16] Not that there was any international traffic—the Soviet Union wanted its own path around the world and access to strategic minerals in the Arctic. In 1991, Russia opened the route to all, but there has never been more than a trickle of traffic, although several big international research programs have looked at its potential.[17]

Now that the ice is thinning, enthusiasm is returning, especially in Iceland where the government has hopes that its ports will be at one end of a new trans-Arctic highway.[18] What is still lacking, says the AMSA report, is comprehensive "cost-benefit-risk analyses for all potential routes of trans-Arctic shipping."[19] The Institute of the North in Anchorage, Alaska, is the only organization to try to take a long hard look at its economics.[20] Mead Treadwell is a senior fellow of that institute and the chair of the U.S. Arctic Research Commission, which advises the president and Congress. Speaking at a naval conference in Washington, D.C., in 2007, he explained that they had looked at "a shuttle that would link the very active North Pacific and transpacific traffic and the very active North Atlantic and transatlantic traffic." Aker Arctic carried out the study, looking at two ships, one a double-acting ship like those used by Norilsk Nickel but modified to carry 750 standard containers rather than nickel plate, and the other a more powerful big brother, able to carry 5,000 containers and travel farther out to sea where the ice is heavier but the water is deep enough for it to pass.

The study looked at year-round operation, Treadwell explained, between Iceland and an old military base at Adak, in the Aleutian Islands. The 4,900-nautical-mile voyage could be made twenty times a year by the larger ship, without help from icebreakers. The results were encouraging. The cost of moving a container from Japan to Europe through the Suez Canal runs around $1,500. The larger icebreaking ship could shuttle a container across the Arctic for $354 to $526 depending on the prevailing ice conditions. But on top of that are the "fairway" fees charged by Russia (currently set at the high price of around $1,000 per container), the cost of the hub ports at both ends of the Arctic, and the journey onward to ports in Europe that are harder to predict.

The Arctic will always be very cold. Although bulk cargoes, including iron ore, coal, and fertilizer, aren't too much affected by a week below freezing, computers from China, cars from Japan, and the manufactured goods that fill many container ships will not be so happy. The unpredictability of the weather and the ice also make schedule keeping difficult and on-time delivery is critical for many goods.

So far, the study has stimulated interest but not action. I called Treadwell at his office in the Institute of the North in April 2009. "We have several shipping companies citing the study as something they have used as a benchmark in their planning, but no one has stood up and said we are beginning to build ships for a trans-Arctic mission. On the other hand, the economics appear to improve as the ice conditions change and shipyards gain more experience with icebreaking ships," he said. Other factors could come into play. Insurance rates could go down if there were better salvage and search-and-rescue facilities in the Arctic, as well as shipping rules and aids to navigation, as the AMSA report points out.

There is also the question of how a trans-Arctic route would fit into world trade patterns. Currently, giant container ships from Asia carrying manufactured goods travel on regular "port rotations," that is, runs to a sequence of ports in the United States or Europe, dropping off cargo. The hub ports in Asia that they leave from are in a class of their own. Singapore, Shanghai, and Hong Kong handle more than 20 million containers a year. Europe and America have more, smaller ports; Long Beach in California and Rotterdam in Holland each handle less than half that number. The largest container ships now being built carry 13,500 TEU (a "twenty-foot equivalent unit," which is the length of a standard container) up from only 4,000 TEU a couple of decades ago. These giant single-screw ships push prices down because they can carry a huge amount of cargo. A ship crossing the Arctic would have to compete with ships carrying at least twice as many boxes, and although the distance is shorter, pushing through ice burns more fuel. When the smaller ship emerges from the Bering Strait, it still has a long way to go to reach one of those hub ports in Singapore or Shanghai. The economics of scale and positioning of giant hub ports make it hard for an Arctic route to compete in a big way. When the planned widening of the Panama Canal is completed, a round-the-world route for giant container ships will raise the bar even higher.

The Arctic route might be used in the next few decades for small quantities of high-value goods, by warships that need to pass between oceans without delay, or even for transporting nuclear fuel far from the reach of terrorists. Unexpected events could boost the attractiveness of the polar route: a prolonged war in the Middle East that closed the Suez Canal would be one. A big boom is very unlikely.

The zone within northeast Russia and continuing past Vardø is where the busiest commercial areas are going to be in coming decades. That's alongside the boom in Arctic tourism and the cruise ships that are heading farther and farther north as the ice retreats.

If you want to travel to the North Pole on board a Russian nuclear icebreaker, circumnavigate Svalbard or Wrangel Island or run the length of the Northwest Passage on a small expeditionary ship, or travel up and down the coast of Greenland or Alaska in a ship carrying a couple of thousand tourists and a choice of restaurants, you are just one click from a shiny brochure arriving at your door.

Back at Vardø, they can see what is happening in Svalbard. "Yes, we are very afraid of this area because very large cruise ships are going up there alone," says Sveinungsen. "Many of the captains know about the Caribbean, but they don't know this ice area, so that's not so good. We are following these vessels and try to inform them about the dangerous areas. Many of these charts up there are more than a hundred years old. The ships are large: the biggest of them carry one thousand people there. We have the coast guard patrol vessel *Svalbard* up there and we have a few helicopters. That's all. So if there is a grounding or a big fire, it will be a tragedy."

There has never been a really large catastrophe or oil spill up among the Arctic ice (the Exxon Valdez catastrophe was farther south), but when I talked to George Edwardson, the Inupiat leader from Barrow, he told me a horror story from his hometown that made the potential consequences clear. "In 1942, we had a U.S. Navy supply boat run aground in a storm, a little over sixty miles east of us in Barrow," he said. "In order to save that ship, the captain had to dump his bunker oil, it went into our barrier islands, and it killed the whole system right there. Pretty close to fifty years later the same storm that put that ship in trouble came back again. The waves washed up where that bunker oil was deposited and that bunker oil started killing that inlet all over again. Up here it's too cold for the oil to vaporize and the ground is permafrost. The toxicity does not go with time." Under threat are the "seals, the whales, the fish, the ducks, and the birds," says Edwardson.

That particular spill has not been studied, but the behavior of the oil that Edwardson describes is familiar. Oil buried in the cold can keep its toxicity for many decades and hit the environment again and again every time it is washed back out into the sea. If there is a spill, tackling it very quickly is essential. There has been little success in dealing with the Arctic spills that have occurred. A 2008 U.S. National Academy of Science's report looked at twenty years of accidents in the Aleutian Islands, including two big oil spills from wrecks and noted that "almost no oil had been recovered and in many cases, weather and other conditions have prevented any response at all."[21]

One of those wrecks was the 740-foot Malaysian-registered *Selendang Ayu*, carrying a cargo of soybeans from Seattle to China on the great circle route that crosses the Pacific in an arc, touching the fringes of the Arctic just north of Dutch Harbor in the Aleutian Islands. The tragic story of its loss provides a warning for any ship voyaging into remote northern waters.[22] In November 2004, the ship was more than fifty miles from land when its engine failed and its master called for help. A U.S. Coast Guard cutter reached the ship six hours later, followed by one tugboat which got a line aboard but could not turn the ship, drifting broadside in twenty-five-foot waves and sixty-mile-an-hour winds. A second tugboat arrived but had to wait until daylight before trying to put another line aboard. Before then, the first towline broke. The captain dropped an anchor which caught and almost stopped the ship. But fifteen minutes later the anchor slipped and the weather worsened. Most of the crew were evacuated by helicopter, leaving eight on board trying desperately to restart the engine. As darkness approached, the ship hit bottom and a final rescue began. One helicopter lowered a rescue swimmer onto the ship to help the evacuation and began pulling the crew on board. Then a huge rogue wave hit the ship, sending spray up into the helicopter's engines and bringing it crashing down into the sea where it overturned and sank. A second helicopter pulled the flight crew and one survivor from the sea; six others were lost. The *Selendang Ayu* broke in half on the rocks, with the ship's master and the coast guard rescue swimmer still onboard. Two hours later the helicopter was able to take them off. Nearly sixty hours had passed since the engine failure. Six men died, 1.7 million liters of fuel oil leaked from the wreck, and over the next months the oiled bodies of over a thousand sea birds and many sea otters were found.

Environmental groups are well aware of the issues. In 2008 the World Wildlife Fund (WWF) called for a moratorium on further oil exploration in the Arctic until effective oil spill technologies are available and deployed, a call they repeated on the twentieth anniversary of the Exxon Valdez spill in 2009.[23] In Alaska, a coalition of environmental groups is calling for a ban on offshore drilling until oil spill measures are addressed. Techniques for tackling oil spills in ice are developing. Ice can sometimes even help contain a spill. In a cold environment, oil does not flow quickly and can be trapped in calm waters among ice floes. The toughest situation, says Ian Buist, an oil spill expert with the consulting firm SL Ross Environmental Research in Ottawa, is a spill in close pack ice with big floes where oil is released under the ice (a blowout from a well would be an example). "It is really hard to find exactly where the oil is," he explains. "You can't find it exactly, so you have to track the ice floes and wait until spring." When the oil hits the cold undersurface of the ice it stops spreading and is gradually encapsulated by growing ice. When the spring melt comes around and channels open in the ice, the oil floats up and appears in melt ponds on the surface. Then there is a chance to burn it off. Helicopters carrying a "helitorch" which fires blobs of burning gasoline jelly (napalm) at the surface can get the oil burning. "You'll probably get two-thirds of it," says Buist. Other techniques can be used. Ships equipped with "skimmers" that scoop up the top surface of the sea and then separate oil from water have been developed to work among loose ice but suffer from an "encounter rate" problem says Buist. Out on open water, skimmers can use booms a few hundred meters wide and pick up a lot of oil quickly. Maneuvering among ice floes makes it hard to encounter the oil at a fast enough rate. Cold water chemical dispersants have been developed, too, but lack extensive testing.

The prospect of leaving a big spill of oil out all winter and using napalm to set it alight in an ugly plume of thick smoke when it surfaces in spring will fill most people with horror, especially as the oil may travel long distances and cross international boundaries. No surprise then that

Shell arranged for a flotilla of fourteen rapid oil spill response vessels for its planned work in Alaska.[24] Open water spills could be tackled efficiently by tugs equipped with long booms and skimmers, although if ice were present, burning would be the best option. Around a fixed well, an oil company can be well prepared. No one can be really ready for a tanker wreck in heavy ice.

One way of reducing risk from spills is to make sure ships don't go where they will do the most harm. Kees Camphuysen of the Royal Netherlands Institute for Sea Research analyzed oil spills in Europe over recent decades and found a surprising result: the smallest spill (600 tonnes from the tanker *Stylis* in 1980) killed the most birds (450,000) while one of the largest (223,000 tonnes from the *Amoco Cadiz* in 1978) killed fewer than 5,000 birds. "It is not the amount of oil spilled that matters most, but rather when and where it happens," Camphuysen writes.[25]

Around the Arctic are special places where spills could have terrifying consequences. Animals with feathers or fur are most immediately vulnerable, especially when they gather in big groups. Seabirds come together at breeding colonies and in polynyas (areas that are open every year among the ice) where they gather to feed before breeding, or at staging areas before migrating. Seals aggregate to pup on the ice. The all-important Arctic cod spawns in the winter under the ice and its eggs have a long incubation time. Biologists are busy making maps of the places in the Arctic where the risks are greatest, but routing ships away from them requires political action.[26]

Oil spills alone do not exhaust the list of worries for the Arctic from increased shipping. There's also the risk that invasive alien species will enter the Arctic in ballast water, and that toxic chemicals used in antifouling paints will harm wildlife. Garbage dumped at sea can pollute the Arctic for years, and the fine, black particles emitted as exhaust from the ship's engines can land on the ice, turn it gray, and speed its melt.

Testifying to the U.S. House Committee on Foreign Affairs in March 2009, Treadwell stated that the Arctic Ocean today is "open to ships of any nation, whether or not those ships are properly prepared for Arctic ice conditions, properly defended against oil spills, or properly served by navigation and search-and-rescue infrastructure. Rules to prevent harmful interactions with marine mammals, or subsistence hunting, are not in place."[27]

We have already seen that many of these issues can be dealt with in principle under the Law of the Sea and by International Maritime Organization conventions, but action is needed to strengthen and enforce them, along with all the other wider issues of Arctic governance. Then there are the issues of ports, charts, rescue tugs, and training of mariners that must be tackled.

When you consider the risks of loss of life or a catastrophic oil spill, you might think that there would be an immediate rush to ensure Arctic safety and to strengthen many of the rules that already exist. History teaches otherwise. The first international conventions on safety of life at sea were agreed upon after the *Titanic* sank; MARPOL was precipitated by the huge oil spill from the *Amoco Cadiz* off the coast of Brittany in 1978; the *Exxon Valdez* disaster drove the switch to safer double-hulled tankers, but it took further accidents to move the deadline from 2015 to 2010. Much could be done quickly, but the pace of change in the Arctic may prove too fast for international organizations. Vardø will be organizing the rescue of a cruise ship in Svalbard before the world demands action.

FINALE

Chapter Fifteen

THE ARCTIC'S REVENGE

It is midnight. The summer sun dips briefly below the hills that lead up to the Greenland ice sheet and the air grows chill. Across the bay, the enormous Eqi glacier is now half hidden in the twilight, but its restless reverberations fill the air; huge cliffs of ice collapse into the sea, calving off bergs that slowly drift away.

After a few nights camped out here, I am beginning to recognize the glacier's range of sounds and rhythms. In the background there are long, deep rumbles that reverberate as if emerging from deep inside a vast cavern. Nearer by, loud crashes and cracks unexpectedly fracture the more comforting bass, sometimes swiftly gaining pace, merging into a crescendo of explosions and ending in angry thunder. There is another sound, too, deceptive in its gentle beginning—a long, soft sigh that often fades ever so slowly into the air, but sometimes can turn violent, gathering strength and rising to a roaring torrent.

I sit and watch carefully, and the glacier's language begins to make sense. The deep rumbles come from huge collapses within the glacier, echoing from high up, where the ice is broken into mile after mile of twisted ridges and crevasses. The long, soft sighs are the signature of snow avalanches, some beginning as a trickle and dying away, others sending larger and larger slopes of snow into a torrent of flowing powder. The crashes and cracks come when the cliffs of the glacier front begin to break up and slide into the sea. Sometimes the collapse spreads, each falling cliff bringing down its neighbor. Soon after, a huge wave breaks on the beach below, arriving with great speed and surprising stealth. I've been warned to walk carefully on the beach because the silent swell can easily outrun you. Unwary people have been swept away and drowned.

Listen a little longer to the glacier and something odd emerges about the pattern of its sounds: it is impossible to guess whether a sound will grow or fade. Sitting there on the Greenland shore, I wonder if the glacier is trying to teach me a lesson. Its unpredictable nature and sudden surprises chime with everything I am learning about how the Arctic is reacting to the changes we've wrought. I call it the "Arctic's Revenge." On land, on sea, in the air—everywhere I hear that the changing Arctic is, in turn, beginning to change the rest of the world. Some of those changes are big, some small, some quick, some slow. Many will be unpredictable and some of them, once they start, will be impossible to stop (although I have met a couple of eccentric Russians who want to put up a fight).

There are so many ways the Arctic might react, and so many different interlinked possibilities, that it scarcely matters where we begin our investigation. I'll start with the wider consequences of the loss of the sea ice, as we can see them taking shape already.

It was always inevitable that if the Arctic lost its summer sea ice, the heat balance of the world would change. Instead of sunlight being reflected back into space from that shining dome at the top of the world, dark seas would soak up the summer sunlight.

Julienne Stroeve of the National Snow and Ice Data Center in Boulder, Colorado, who I'd earlier talked to about predicting changes in the ice, has a brightly colored map of the Arctic Ocean that shows what is happening. On the map you see a great red blob where, over the last four years, temperatures at the end of the summer have been around 3°C higher than the average in the previous twenty years. Within that large red area are smaller, brighter spots where temperatures have risen around 5°C. The increases are worrying, but it is their position that is really important. "The warming is directly located over those areas where we've lost all the ice," Stroeve explains.[1] The Arctic amplification of air temperatures driven by the ice–albedo feedback has arrived. Instead of bright, white ice reflecting sunlight back into space, the growing area of dark open water is soaking it up.

Back in 2007, the scientists of the Intergovernmental Panel on Climate Change (IPCC) predicted that global mean temperatures would rise by 2.5°C to 3°C by the end of the century. In the Arctic the rise may be more like 7°C. The white cap of ice has always helped cool the planet, but its powers are weakening. "The Arctic is really the air conditioner of the northern hemisphere, and as you lose the sea ice you change the air conditioner, and the rest of the system has to respond," explains Stroeve. "You start affecting the temperature gradient between the Arctic and equator which in turn affects atmospheric patterns and precipitation patterns." So what will happen next as the Arctic warms? "Our research is in its infancy," says Stroeve. "Exactly how this is going to play out, we really don't know yet." Already the first subtle signs of the Arctic's reach are appearing. When there is little ice in the Arctic in summer, there is less rain in winter across the United States and Scandinavia and more rain in the north Mediterranean and Japan.[2]

A little nearer to the ice, the impact of the Arctic amplification is clearer. The pool of warm autumn air over the Arctic seas is spreading out over the surrounding Arctic lands and is warming them, too. The big surprise is in how far this widening gyre of Arctic warmth can reach and what it may do. Stroeve's map of the Arctic, with its bright red hot spots over the areas of open seas where temperatures are soaring, fits seamlessly with another map of the wider Arctic from her colleague, computer modeler David Lawrence.

On his map,[3] the coastal strip of the Arctic is painted deep magenta, with various shades of red continuing far inland and across the top of Greenland. The colors denote zones where temperatures are predicted to rise at rates of 3°C and 2°C per decade as a result of the warming sea: they extend over 900 miles from the Arctic seas, deep into Canada and the tundra of Siberia. All of the north seems to have caught fire. "It's not just warming the land at the margin of the Arctic seas, the whole of the pan-Arctic land is being heated up," explains Lawrence. He estimates that the rate of warming on land set in motion by rapid ice loss will be some three and half times greater than that predicted from current long-term trends.

Already the temperature on land is changing around the Arctic and its impact is growing. Donald "Skip" Walker of the University of Alaska at Fairbanks is a geobotanist who has been measuring how "green" the Arctic is, combining both satellite observations that pick out green vegetation from space and studies on the ground. I heard him speak at the 2008 American Geophysical Union conference in San Francisco where he explained that between 1982 and 2007, as the summer sea ice declined by a little over a quarter along a thirty-mile band off the Arctic shores, the greening index along the North American coast leapt by almost a quarter. Where once there was bare tundra, shrubs are marching toward the sea. Trees will follow. As they advance, some beautiful and rare desert-like ecosystems of the farthest north, where the Arctic poppy grows, will be pushed off the edge of the map and disappear forever. A little farther south, the Arctic's characteristic tundra is under siege. That tundra forms a strip around the Arctic's sea, equal to the area of the United States, but rarely more than a hundred miles wide.

I've seen these places and for me their gradual disappearance is a profound loss, but a bigger worry is that a greener Arctic is a darker Arctic. Trees are darker than shrubs, which are darker than tundra.[4] And trees and shrubs shed snow. Fly over an expanse of tundra in winter and it is almost pure white; a forest is not. The snow is lying on the ground beneath the branches of the trees, acting as a blanket that helps keep the soil warmer than usual in winter, speeding its thaw in summer, while the darker vegetation above soaks up heat that would normally be reflected back by the white tundra into space. The change that begins with the loss of ice widens and the earth grows warmer still. At the end of his talk, I asked Walker if there was any upside to this change. Surely, more vegetation would mean more carbon will be taken out of the atmosphere? "Yes," he said, "but the amount is small."[5]

The spreading Arctic warmth is being felt in the north of Greenland already. In 2008, Marco Tedesco of the City College of New York reported in the journal *Eos* that the northern part of the Greenland ice sheet showed a record snowmelt.[6] The melt lasted eighteen days longer than the previous maximum, which may not seem significant until you learn that it pushed the melt season from its usual twelve to fifteen days to thirty-five days. "High surface temperatures are most likely to blame," wrote Tedesco. "The average temperature here is going up by 1.5°C every decade."

Maybe it is not surprising that an eleven-square-mile chunk of the Petermann Glacier broke away nearby. A huge crack opened up farther back in the glacier and another twenty-three square miles were under threat. A few weeks afterwards, across the water on Ellesmere Island, the 4,500-year-old, nineteen-square-mile Markham Ice shelf broke off and drifted away into the Arctic sea. Tedesco ran off this list of disasters when I met him at a conference and asked him what he had seen in Greenland, and then he added with a wry smile, "It is worrying when you see so many extreme events in a row."

Farther south in Greenland, temperatures are also rising. Tedesco told me that he had also picked up a record snow melt on the ice cap there in 2007.[7] Here we join a bigger and much better-known story about the way the changing Arctic may change the world. The Greenland ice cap is melting away at a speed that nobody predicted in the past. The melt might be partly due to a natural cycle that will slow or reverse, but if it is not, then sea levels around the world are going to rise faster than anyone has seen before.

The fastest changes are happening just thirty miles south from where I sat listening to my restless glacier. Ten huge ice streams drain the enormous mass of Greenland's ice cap and the Jakobshavn Glacier is the greatest of them all. Almost two miles wide and fed by the ice of 7 percent of Greenland's interior, Jakobshavn has been retreating slowly up the long fjord in which it sits for over a hundred years.[8] In 1992, scientists logged the glacier moving along at a steady 3.5 miles a year. An unexpected spurt followed: in 2000, it was running at 5.8 miles a year, and by 2003, roaring along at 7.8 miles a year. As the glacier moved faster, the ice deeper within the ice cap that fed it began to thin rapidly. Greenland's ice was rushing for the exit, and no one was sure why.

The final outlet for the glacier is just a little south of the town of Ilulissat (formerly Jakobshavn) and is one of the wonders of the world. Here you can sit all day, with the wooden houses of the town neatly painted in reds, blues, yellows, and greens on the hill behind you, and watch huge icebergs sail by across the bay. There are bergs that remind you of medieval castles, complete with crenellations and towers. They have floated unchanged from the glacier with cliffs and crevasses intact. Other giants have capsized and now show their water-polished undersides; they rise from the sea as smooth-faced mountains of ice. Some have capsized many times and you can read their history from the waterlines written across the bergs at a range of angles. Occasionally a berg is marked by a vein of the most brilliant blue, a crack that filled long ago with stream water and froze so clear to sunlight that the true blue color of the ice can shine from it. Look to the south and the icebergs form a jostling wall that sometimes advances menacingly right up to town. This is the mouth of the fjord into which the glacier spills its ice.

There is busy trade for fishing boats carrying tourists as far into the fjord as they dare go. Even then you are still thirty-four miles from the front of the glacier where the bergs calve. They drift through the ice-clogged fjord for months until they reach the sea, where a now drowned glacial moraine forms a shallow sill. Here the larger icebergs run aground and must wait until they grow smaller and can tumble into the deeper sea.

The strange changes that scientists were picking up at Ilulissat were soon seen elsewhere. Back in 2006, a remarkable paper was published in the journal *Science* analyzing data from a satellite survey of Greenland.[9] Its lead author was Eric Rignot, who had started off studying astronomy at the University of Paris but in the 1980s had moved to California to focus on ways to see from space what was happening on Earth. He is now at Caltech and, alarmingly, he found that the glaciers everywhere in the southern part of Greenland were collapsing and that the melt was spreading to the glaciers farther north.

Greenland holds a staggering amount of ice. Up on top of the ice cap, all you can see is an infinite field of white, but under your feet may be as much as two miles of ice, some tens of thousands of years old. You won't be standing two miles up in the air, though, as the ice is so heavy that it has pushed the earth's crust down below the level of the surrounding sea. If all 700,000 cubic miles of it were to melt away, the earth's sea level would rise by around twenty-three feet, causing an unimaginable catastrophe.

A rise of twenty-three feet would engulf large parts of Florida and Bangladesh, drown the Nile Delta, and flood many of the world's big cities. Still, everyone has always assumed that you couldn't possibly melt away such a giant volume of ice in less than a millennium, time enough to build new cities and relocate the 15 percent of the world's population that would find their homes underwater.

That view assumes that the ice just melts slowly away as global temperatures rise, as if it were an ice cube left out to thaw. No allowance was made for "dynamic effects," that is, the rapid discharge of great chunks of ice straight into the sea. The 2007 IPCC report simply excluded fast-moving glaciers, recording that "models used to date do not include the full effects of changes in ice sheet flow because the basis in the literature is lacking." Glaciologists love that quote as it shows that their discoveries are really new.

If you take into account the rapid collapse of the glaciers, how much water is Greenland adding to the world's oceans? In 2008, Rignot teamed up with scientists from around the world and estimated that the ice sheet had been losing 30 gigatons of ice a year from the 1970s through the 1980s, 97 gigatons in 1996, and between 239 and 305 gigatons in 2007. I read their results in the journal *Geophysics Research Letters* and it left me stunned.[10] A gigaton is a billion metric tons, or the weight of a cubic kilometer of water. Add the latest annual figure of 305 gigatons to the oceans and the sea level rises by close to a millimeter. Keep going faster for a century on top of the natural thermal expansion of the oceans as they warm and ice melting elsewhere and that is enough for governments around the world to have to add billions to the cost of coastal defenses. The acceleration is deeply worrying. Its cause appears to be those rapidly moving glaciers: the paper shows that they account for between 40 and 80 percent of the ice loss.

I called up Eric Rignot in his laboratory and asked if he was surprised too. He laughed. "Even just a couple of years ago, to state that the ice sheet was losing as much mass as it is, would make me considered a mad man. I think if you had told people in 1990 that I would make a prediction in 2008 that we are going to lose three hundred gigatons per year of ice in Greenland, everybody would have laughed. He is not serious, they would have said. There is no way you can get anything like that." So what will happen next? "We see acceleration. It's not a linear trend; it's more rapid than that. I don't know where it's going to go. Ten years ago we thought we knew everything. Now we know we don't. Predicting what might happen in the next fifty years with continued warming of the climate is a difficult game. I think the answers might change every year."

I had already had some experience of rapidly changing answers. Not long before talking to Rignot, I had been up onto the Greenland ice cap to see the explanation for the sudden speed-up in the glaciers—or so I naively thought.

An easy way up is to follow the ten-mile-long route laid out by the French scientist Paul Emile Victor. Soon after the end of the Second World War, Victor realized that there were surplus military tracked vehicles to be bought cheaply. He used them to help build a winding route up through the hills not far from where the Eqi Glacier reaches the sea. Some obstacles were dynamited, and on a really tough stretch across a steep valley he built a half-mile-long cable system to shift heavy equipment. The remains of it are still there, along with abandoned sleds, bits and pieces of equipment, and some sun-bleached wooden crates that look as if they once held champagne. It is still the perfect spot for a picnic.

Victor's route is much overgrown now. In early summer, the tundra is green with moss, lichens, and dwarf willow. Ptarmigan herd their huge families of newborn chicks away from you as you plod up the hill toward them. Sometimes you find one sitting on its nest, utterly still, relying on its dappled feathers to conceal it among the colors of the surrounding tundra. Sometimes it would have been better to flee. Just a few feathers and an abandoned egg lie at another nest. An Arctic fox had visited.

Higher up, the greens and browns give way to the dull, black-and-pewter-colored mounds of the lateral moraine pushed up by the moving ice. Then the ice cap lies in front of you, stretching 600 miles all the way to the east coast of Greenland. At first you can't take your eyes from the infinite whites and blues of the undulating plain of ice and the streams of crystal meltwater that cross it. But I was up there to look for "moulins," or rather, to listen for them.

Moulins are the places where melting water, running in rivers across the surface of the ice, suddenly vanishes down a hole, creating a churning waterfall under the ice which sounds like an old-fashioned mill race. You can hear a moulin a long way off and when you eventually come close you put on a climbing rope, attach the other end to a strong person at a good distance, and approach carefully. The rush of foaming water disappearing into a vent amid the brilliant blue of the ice is spectacular. So too is the slipperiness of the water-polished ice; one mistake and you'll drown in the glacier's plumbing.

I returned from the moulin hunt believing I had seen the cause of the ice cap's slide toward the sea. With rising air temperatures, meltwater was draining way under the ice and lubricating the glacier's slide toward the sea. That new theory had made headlines just a few years earlier.

Tavi Murray set me right. She is a glaciologist from the University of Swansea who I had run into by chance in a little airport in Greenland a year earlier. She was just off to survey thinning glaciers and told me about her plans with great enthusiasm. I called her to say I'd finally been up there myself and seen the moulins. Yes, she said, a few years ago, scientists had discovered that meltwater was getting down to the bedrock at the bottom of the ice and lubricating it, speeding the flow. "That was a big 'wow,'" she said. This potentially catastrophic effect made its way into Al Gore's documentary *An Inconvenient Truth*. Yes, it is still "very interesting," said Murray, but it has turned out to be not quite the full explanation we need. When scientists were on hand to see a huge lake drain down a moulin in 2008, they found the glacier only managed to lurch forward three feet.[11]

So what is making the ice speed up? Rignot thinks the fast glaciers have been "ungrounded." Warm water is getting underneath the ends of glaciers where they stick out into the sea and melting them so they are no longer jammed onto the bottom of the fjord. "It is as if someone was lifting up the part of the glacier which reaches the ocean and so the ice upstream is able to flow faster," he explains.

For many of Greenland's glaciers that's not going to stop for a long time. The Jakobshavn Glacier follows a deep trench below sea level for sixty miles back inland, he says. Once the water starts to lift the glacier, the process just continues.[12] "The take-home message is that if a glacier is grounded deep below the sea level, and reaches the ocean and it starts to retreat, there is very little you can do to stop it," says Rignot.

That was scarcely reassuring, but both Rignot and Andy Shepherd, a glaciologist from the University of Edinburgh, assured me that if I wanted to really worry, I was visiting the wrong pole.[13] The glaciers of the West Antarctic ice cap have also "unplugged." We've got twenty years of observation in West Antarctica, explained Shepherd. "In the first ten years, we lost one hundred cubic kilometers [one hundred gigatons] of ice per year. In the second ten years, we lost two hundred cubic kilometers of ice per year. Keep doubling that to four hundred, eight hundred, and you get a very, very large sea level rise in a very short time. There is no theory to support that, but there is also no evidence that this is slowing down. We just see it getting bigger and bigger."

I feel frustrated that I can't find more certain predictions, but I have to accept that I am at the frontier of knowledge. What we do know is that there is a very large threat (add West Antarctica to Greenland, plus the inevitable expansion of the sea as it warms, and you are looking a rise in global sea levels of sixty feet) which may arrive over thousands of years, but which is showing some signs of arriving sooner rather than later.

To seek a little comfort among all this uncertainty I turned to Tad Pfeffer, professor of engineering at the University of Colorado at Boulder,[14] who has done something very clever. He drew together surveys of the bottom topography of the big glacier outlets around the world and added numbers on how fast the glaciers could go. That provides a good guess of the maximum amount of ice that could come pouring out of them. He told me that the worst case is two and a half feet (and that's a maximum, not a prediction) this century. His result has been controversial. It suggests that some alarming predictions are unjustified but the limit is much higher than the IPCC's now rapidly aging view. If we came close to that limit, that would be pretty serious, especially if you are Dutch. You will need to move to another country.

In the changing Arctic we never quite know how many more surprises are waiting for us. We knew nothing of high-speed Arctic glaciers a few years ago. And until just a couple of years ago, we had no real idea that as the Arctic warmed, methane would begin fizzing out of its innards and into the atmosphere. Methane is a powerful greenhouse gas, but unlike carbon dioxide, it does not stay in the atmosphere very long. It packs a very strong but short punch. Now we know that methane is bubbling out of ponds in the tundra and out of the Arctic's shallow warming seas in places where it had never been seen before.

I had seen gas bubbling out of small lakes on a warm day in Alaska but was not at all sure whether it was methane. Somehow, among the 16,000 scientists who showed up at San Francisco's Moscone Center for the fall 2008 meeting of the American Geophysical Union (a record for one of the largest science conferences in the world), I managed to find Katey Walter, an expert on bubbling lakes from the University of Alaska. It wasn't really so hard to find her as I knew what she looked like; although only a few years out of grad school, she's already published papers in *Nature* and *Science* and I'd seen her on television.[15]

I asked how I would know if I saw methane, expecting a reply along the lines of, "Collect the gas and take it to a laboratory equipped with a mass spectrograph." Instead, she gave me instructions: "Hold out a lighted match. If it's 10 percent methane or more it will burn. But be careful. I've seen eyebrows go. I've seen a big piece of my hair fly off and burn."

Walter is an unusual person, with a love of the outdoors and Russia. She explained that she went to Russia as a high-school exchange student, learned the language, and became "very passionate" about the nation. When a chance came to go from Alaska to work in Siberia at the Northeast Science Station in Cherskii, she grabbed it. Cherskii is in the furthest reaches of Yakutia and about as remote a spot as you can find on the planet. It is also very cold; mean January temperature is -38°C. Winter walks on frozen Siberian lakes helped Walter to answer that question of how much methane was bubbling from lakes. "The surface cover of ice trapped all the gases," she explains. "It's transparent so I could walk across the ice and see all these hot spots of methane bubbling. I knew how many there were and how strong they were." With the ice to help her find the methane, she could use an umbrella-shaped trap to measure their power. "It was very simple but it worked," Walter says.

The amount of methane proved far greater than anyone expected—and once again it hadn't been included in any of the climate models. Plus, the lakes don't stand still. "The lake just grows," explains Walter. "The Russian word for it is 'eats the permafrost.' And I love it because it's right. If you look along the edges of these lakes they are not nice and smooth. They are rough, as if someone is taking bites out of a cookie. The lake water is warmer and it migrates across the surface by eating the permafrost, digesting the materials and burping out the methane."

The methane is created when permafrost—the frozen ground just beneath the surface—melts, allowing microorganisms to break down its rich store of organic material. "More methane increases global atmospheric temperatures and causes more permafrost to thaw," says Walter.

With any discovery like this, two questions immediately leap to mind: has anything similar happened before in Earth's long history that might give us clues about what will happen next, and how big the impact might be? Walter looked back to the "scars of old lake basins" that cover the Arctic, worked out estimates of when they formed, and found they corresponded with a surge in methane levels at a time when the earth was warming rapidly 14,000 years ago. Arctic lake methane might have been an important trigger in past climate change alongside methane from expanding wetlands.[16]

She has since expanded her studies of lakes to North America in addition to Siberia. Currently, Walter says, the permafrost lakes add 25 teragrams of methane to the atmosphere. "It isn't a huge number but it is a large number, given that the total methane entering the atmosphere from everywhere in the world is about 560 teragrams. And it's a source that hasn't been accounted for yet. If the future pattern of lake thaw is consistent with what happened in the past, then permafrost thaw lakes could be adding 50 teragrams of methane a year if they thawed over thousands of years, or 500 teragrams a year if they thawed over hundreds of years." Which will it be? "Our understanding of how these lakes form and grow says it is a centuries timescale," says Walter. "Although, if you really warm permafrost and start it going, it could go quickly. So I don't know for certain. Is it one hundred years or one thousand years?" She leaves the question with me.

I quickly find I have a bigger lake to worry about when I meet Igor Semiletov, an oceanographer from the Russian Academy of Sciences. He is also at the giant conference and has recently returned from a series of ocean cruises off the coast of Siberia with a troubling tale. "We found huge bubble clouds of methane in water columns in the Laptev Sea and the East Siberian Sea," he explains. "Permafrost lies under 80 percent of the entire area; we didn't know that this huge carbon pool is extremely vulnerable."

These shallow seas have formed quite recently. Ancient permafrost, once part of the land, is now under the sea and may come into contact with water that can thaw it quickly. With more than 75 percent of the area under less than 130 feet of water, says Semiletov, escaping methane can bubble out into the atmosphere before it has time to dissolve in the water. I ask him, "When did this start?" He didn't see anything like this on trips in the 1990s, he says, and thinks that "the subsea permafrost is in a transition phase now." More worryingly, Semiletov adds that the permafrost is "failing to seal ancient carbon, which includes methane hydrates and natural gas."

A conversation with Semiletov feels a bit like being passed a series of hand grenades. You are just praying that no one has pulled the pin out of any of them. Methane hydrates are Earth's real climate bomb.[17] I have a little indirect experience with them. Up in the Shetland Islands, north of Scotland, there is a quiet little bay I like to walk around. Heading inland a little, you'll find something very odd. Streams cut though the ancient peat, and, high up in a dark rill, there is a thin layer of what looks like sand sandwiched between layers of the black peat.

A professor of geography at the University of Oxford called David Smith has looked at this sand very closely. "It was laid down when a tidal wave hit Shetland around eight thousand years ago," he says. From the gradation of the sand particles he could work out how it settled and how deep the water must have been above it. The answer is that the wave washed up water at least a hundred feet deep. That tsunami was caused by a huge underwater landslide on a steep shelf off the coast of Norway. One candidate for triggering that landslide is the warming of methane hydrates.

The seabed in that region of Norway is rich with hydrates, a frozen form of methane in which the gas is trapped in an icy cage. It will only be released if it is warmed or if the pressure falls. Around the time that the tsunami hit, the sea had warmed. Did it trigger a sudden release of methane which in turn, triggered the slide? The idea is controversial but it does tell you something about the nature of the seabed methane hydrates. Left alone, they are quite safe. But if they are warmed, they may be released in a sudden burst. And there are a lot of them: if just 10 percent of the world's methane hydrates were released over a few years, it would be the equivalent of increasing atmospheric carbon dioxide concentrations tenfold. The impact would be catastrophic.

I ask Semiletov if there is a real risk of releasing methane hydrates in the area he has visited. "The only thing stopping gas hydrates being disturbed and released to the atmosphere is the existence of subsea permafrost. If the subsea permafrost fails, all this hydrate would be released. If that happens, the world will be changed. There are at least five hundred gigatons of hydrates stored in the Siberian Arctic shelf area. In the atmosphere, we have less than five gigatons of methane." Is it hundreds or thousands of years away? "I don't know," says Semiletov, "we do not have enough data. Too few people have looked under the sea."

Semiletov has handed me another hand grenade. On land, the state of the permafrost is clearer, and attempts to predict its future are more mainstream. The best person to provide a snapshot of what is happening is Vladimir Romanovsky, born in Siberia but now at the University of Alaska. I guess that his Siberian past led him to the permafrost. Romanovsky laughs and explains that he trained as a marine geophysicist in Moscow but he found that he "suffers very severely from sea sickness," and so turned back to the solidity of frozen ground.

Romanovsky is currently running an international project to coordinate a network of permafrost observatories around the world. Each observatory is quite simple, he explains, just a one-inch-diameter borehole that typically drills a few dozen meters into the ground but in extreme cases can go a kilometer deep. Temperature sensors record what happens inside. With 400 boreholes in operation, a pretty consistent picture is emerging. "Most of the sites are showing increasing temperature over the last twenty to thirty years. Siberia is now a kind of hot spot, and Alaska was a hot spot during the 1990s when there was a major increase in permafrost temperatures. In West Siberia, we are starting to see thawing of the permafrost. And in places like the interior of Alaska, for example, ice is being lost inside the permafrost. At some point, of course, it will start to thaw."

I wonder why the warming has moved to Siberia and quickly find that I have traveled in a circle and returned to where I began with the loss of the sea ice. "Most of the Siberian coast and inland is influenced by the Arctic Ocean," says Romanovksy. "The dramatic decrease in sea ice really has this very big impact on climate." I am back with those models of the spreading Arctic Ocean warmth and this time see their impact on the frozen ground rather than the vegetation. Perhaps not too surprisingly, NCAR's David Lawrence, who made those models of the spreading Arctic warmth, is now working with Romanovsky to figure out how quickly the permafrost might change.[18]

It's not easy, Lawrence explained to me. His first attempt, which concentrated only on the top layer (the first 11.5 feet) of frozen ground, created a major shock in 2005.[19] That model showed that we could expect 90 percent of the top layer of permafrost to go this century. The current 3.9 million square miles of continuous permafrost that rims the Arctic (farther south there is a zone of patchy discontinuous permafrost) would shrink to a mere 390,000 square miles.

That model was quickly criticized by permafrost researchers, including Romanovksy, for having left out some key details. Now they are all working together and have come up with a more realistic picture.[20] Still, it left no room for celebration: the new figure is that 80 percent of that top layer of permafrost will go by 2100.

There are many reasons to find this deeply troubling. Within the Arctic lands, the ecosystems and the plant and animal life that are adapted to frozen ground will be utterly transformed. The way water runs off the land and down the rivers of the Arctic will change in ways we cannot predict, while the rest of the world must fear the long-term impacts as feedback loops that we don't really understand are set in motion. Warming permafrost will begin to turn its immense carbon stores into methane and carbon dioxide.

In 2006, Sergey Zimov, a Russian researcher who set up the Northeast Science Station in Cherskii, together with colleagues from the University of Alaska, published a paper in the journal *Science* estimating how much carbon might be frozen away in the Arctic and comparing it with the carbon found elsewhere.[21] Up in the atmosphere there are 730 gigatons of carbon, most of it in the form of carbon dioxide. That amount has grown from the 560 gigatons of preindustrial times. We are adding 6.5 gigatons each year by burning fuels from the earth's ancient fossil store. In the permafrost, Zimov estimated there was another 950 gigatons of carbon. If it were released, then greenhouse gas concentrations in the atmosphere would more than double, taking us back many millions of years to the era when temperatures were more than 7 degrees higher than they are now.[22]

Of course there is no way that the carbon will all be released at once. I suspect that the Arctic prefers a long, slow, sweet revenge. The permafrost will melt over hundreds and thousands of years. The very top layer will go quite quickly. Lakes that eat into the melting permafrost will bubble out methane ever faster throughout the century. The rotting ground will produce carbon dioxide. The slowly warming shelf seas will bubble methane and raise endless worries about the stability of those huge stores of methane hydrates at their bottom. There will be no sudden disaster but a long, slow, rising trickle of greenhouse gases and a progressive tightening of the noose. As greenhouse gas emissions rise from the Arctic, the only choice we'll have to prevent still faster warming will be to reduce our own emissions. Year by year, the targets we'll need to reach will grow tougher.

There is much more that could be said, for we have still hardly scratched the surface of all the ways that the changing Arctic will change the rest of the world. The melting ice and swelling rivers will send pulses of freshwater into the Atlantic and that will slow the warm ocean currents coming up from the south over this century. A huge store of freshwater floating on the saltier sea is now building up in the Arctic. Another large group of researchers are worrying about when it may go. Much less noticed is that a steady, extra trickle of fresher water out of the Arctic has begun to change the ecosystems of the North American Atlantic coast. Overfishing off the coast of Newfoundland destroyed the cod there, but the arrival of the Arctic freshwater is helping prevent their return.[23]

The sensible thing to do is to try to control human emissions of greenhouse gases as quickly as possible. But there are other things too. I take inspiration from something that is happening in Siberia. It may be quixotic, it may fail, but its vision is grand, and that should be enough to inspire us all to buckle down and get some global action to bring greenhouse gas emissions down. It is a project that tries to tackle the melting permafrost by undoing a change wrought by humans thousands of years ago.

The Siberian tundra is particularly rich in carbon because it contains a special soil, *yedoma*. When much of the northern world was covered in glaciers, the lands of eastern Siberia remained free from ice and accumulated wind-blown loess that supported a rich, savanna-like grassland. In that era, these vast grasslands were home to mammoth, bison, horses, musk oxen, Siberian antelope, rhinoceroses, and Siberian tigers. The closest thing left on the planet is the savannas of Africa and their herds of large mammals. The grasslands must have been a wonderful sight.

The large mammals of Siberia did not just live in this grassland ecosystem, they also created and maintained it. Trampling hooves stop mosses from growing and encourage grass instead. Grass uses up a lot of water and dries out the soil. Herbivores eat the grass, digest it, and recycle nutrients to the soil quickly. That keeps the grass growing and develops a thick soil which is a rich sink for carbon. Without herbivores, the ecosystem shifts to a soggy tundra where mosses trap moisture in the ground. With the right number of herbivores the ecosystem moves toward grassland and stays there.

Why did this ecosystem of mammoths and tigers vanish? Sergei Zimov believes that the arrival of human hunters was to blame.[24] They wiped out the large mammals, just as they have done in so many other places. When that happened, the steppe stopped being a rich sink for carbon, where grass and bones were building up a thick soil. Instead, that store of carbon froze up—and now it is waiting to be released back into the atmosphere.

Zimov's solution is to turn back the clock and make the tundra into grassland again. He is working to create what he calls "Pleistocene Park." He has fenced off sixty square miles of Siberian tundra and has begun stocking it with reindeer, horse, and moose. If he can trigger the move toward grassland, he can later introduce bison, musk ox, and even fierce predators.

I'd long wanted to hear more about Sergei Zimov's project so I was pleased to find him at the AGU meeting in San Francisco standing by a poster exhibit on the Siberian permafrost. I thought he looked astonishingly young for such a well-known scientist and I said so. "I am Nikita," was the reply. "You are thinking of my father." The young man pointed to a large, bushy-haired, bearded gentleman standing nearby who was carrying on a rapid conversation with three people at once.

Sergei and Nikita work together on the project. Nikita explained to me that there would be many advantages to restoring the grassland. "You know the problem," he said. "There are five hundred gigatons of carbon in the Siberian yedoma. It took thirty to forty thousand years to accumulate. We really don't want to release it." Returning the ecosystem to grassland would turn it back into a carbon sink again. Most important is that herds of animals will trample the snow. Fallen snow is a great warm blanket for the ground. "The air can be -40°C," says Zimov, "but under the snow it is only -5°C to -10°C. When the snow is trampled that changes completely and the soil surface is at -30°C. That keeps the permafrost cold and keeps its carbon safe," he says.

The ultimate dream is to bring back the mammoth too. Mammoth bones are found easily in the tundra, and local indigenous people make a living by collecting and selling them. The Zimovs have brought Japanese scientists here who have recently found ways to resurrect DNA from dead, frozen mice. The return of the mammoth is not impossible.

I asked Katey Walter if the restoration of the Siberian grassland might really be practical. "I think it's practical," she replied. "Not many people are willing to take such risks and create a huge new ecosystem. That's one of the wonderful things about being in the middle of nowhere in Siberia. You can do it."

Could mammoths really save the tundra and help save the world? I am not convinced, but it is such a wonderful idea that it might call us all to action. And there is still one last thing that could be done.

Chapter Sixteen

BLACK AND WHITE

Traveling around the Arctic, there was one man I kept running into again and again. I came across him in Greenland, Svalbard, Norway, and Finland and his name recurred on old maps of Siberia, and even on the planet Mars. I can't say that I met him in person, as he has been dead for more than a hundred years, but I do have a good idea of what he looked like—a firm gaze, jutting jaw, a thick head of hair swept back from his forehead, and a luxuriant mustache. A bust of him stands in Kaivopuisto Park in Helsinki from where he looks out across the gray Baltic Sea as the sailing boats go by. Alongside him are several stone panels depicting his own boat, the *Vega*, a 150-foot, three-masted whaler equipped with a powerful steam engine, and the route he took in it around the northern coast of Siberia.

He is the nineteenth-century scientist and explorer Baron Adolf Erik Nordenskiöld, the first man to travel the full length of the Northeast Passage.[1] He left Tromsø in Norway in 1878 and sailed east, spent one winter locked in the ice not far from the Bering Strait, and arrived at Yokohama in Japan in 1880. Somewhere toward the end of his journey he must have crossed tracks with the U.S. ship *Jeanette*, sailing off in the other direction, toward the North Pole and the disaster that was to inspire Fridtjof Nansen. The recently opened Suez Canal gave Nordenskiöld a faster route back home to Sweden. Although Nordenskiöld was born here in Helsinki, he was banished by Finland's Russian rulers for having made an excessively patriotic speech and lived most of his life in Stockholm. Both Finland and Sweden now claim him as a national hero.

Nordenskiöld is important to me not for his great voyage but because

245

of one of his other journeys. I learned about that trip only by chance, but it led me to hear about one remarkable and little-considered way that we might yet slow the Arctic melt. It is not a solution to global warming, but it might help buy time for the planet and the ice while we find a way to cut greenhouse gas emissions. None of the scientists who helped me dissect the causes of the great ice loss mentioned it, so I am grateful that the long-departed baron's adventures came echoing down the centuries to me. They carried me to another group of scientists and a handful of activists with an optimistic call for action in the Arctic.

The very first time I heard of Nordenskiöld was when I climbed the mountain named after him in Svalbard, in a bitter wind. A sheet of ice built up rapidly inside my high-tech windproof climbing jacket as we approached the summit. ("Useless for the Arctic," my guide said contemptuously. "If you want to stay alive you should wear fur." No doubt Nordenskiöld did.) A little later, I passed Nordenskiöld Land and then the Nordenskiöld Peninsula, all in Svalbard. On old maps of the Arctic I'd also found the Nordenskiöld Sea (now the Kara Sea) and the Nordenskiöld Archipelago. There is a nice medium-sized crater on Mars named in his honor, too.

Nordenskiöld made nine expeditions to the Arctic. As far as I know, the only place he ever failed to reach was the North Pole. He did try, but he made a mistake in choosing reindeer to haul his sleds rather than dogs. The reindeer ran away. Later in life he settled down and spent his time collecting maps. His vast collection is now in the University of Helsinki. It is the last of his expeditions—to Greenland in 1883—which led me to think again about the melting ice. I learned about it only after visiting an area just a few miles from where he had been 125 years earlier.

I had hiked up to the Greenland ice cap to look at moulins, those big holes where a churning stream of meltwater plunges deep down under the ice. Before I found the first of them I came across something odd. Here and there, puncturing the ice around my feet were scores of perfectly round holes. Many were an inch or so across and four or five inches deep. Some were smaller. Others were much larger and partly filled with water. Rising from the bottoms of these ponds were pointed islands of ice, resembling a chain of miniature volcanoes.

I had never seen these holes before, but it was not too hard to work out how they came about. At the bottom of any hole, small or large, there was always a thin layer of dust. Each hole begins when some black dust, often blown from the nearby moraine, lands on the pure white ice. The low-albedo dust soaks up the sunlight, warms, and begins to melt its way downward, leaving a smooth, round hole as it goes. Where many holes run close together, their walls partially melt into each other, leaving those little pointed islands of ice in the middle of a larger pond.

I discovered later that the holes are called "cryoconite" or "ice dust" holes. The man who had named them was, of course, none other than Baron Nordenskiöld. He had found the holes a great nuisance and grew quite angry with them. During the night a thin skin of ice would form on the surface of the bigger holes, making them hard to spot in the morning. "It was impossible not to stumble into them at every moment," he complained in an account of his journey. "They are more dangerous than crevasses. They lie, with a diameter just large enough to hold the foot, as close to one another as the stumps of the trees in a felled forest.[2]"

He knew at once how the holes were created, but his curiosity and knowledge ran deeper than mine. Nordenskiöld had studied both chemistry and geology and at the age of twenty-six had become superintendent of the Mineralogical Department of the Swedish Royal Museum. He was excited by a layer of a fine dark substance at the bottom of the holes that seemed different from mere wind-blown sand and tested it as best he could. The substance could be "drawn to a magnet," he found, and when it was ground in a crucible and heated with a blow pipe, it burnt with colors that suggested it contained zinc and iron. He recorded his results in the recently launched U.S. journal *Science* in 1883.

Nordenskiöld didn't really know where this sediment came from, but we have a very good idea now. Timothy Garrett, a professor of meteorology at the University of Utah, reexamined the 1883 paper and thinks the dust was a by-product of smelting and coal burning during the late nineteenth-century industrial boom in Europe and North America. That early industrial pollution had been carried right up to the Arctic and dumped on the snow where it began to soak up heat and melt it away. Garrett published an article on early pollution in the Arctic which I came across as I searched for information about Nordenskiöld.[3]

It is possible to find evidence of that old pollution today by drilling deep into the west Greenland ice cap and pulling out frozen cores of ice several hundred meters long. Deep in the core is the compacted snow that fell hundreds of years ago. With skill and care, the layers of ice in the core can be read and dated. Joseph McConnell of the Desert Research Institute in Reno, along with Charles Zender of the University of California at Irvine and colleagues looked in these cores for a range of air pollutants. They published the work in the very same journal, *Science*.[4] Sure enough the ice record showed that Nordenskiöld had been there just when the pollutants drifting to the Arctic from the burning of coal were coming to a peak.

Garrett isn't a historian of science, and when I tracked him down at the University of Lille in France, he explained that he had looked back at old accounts of the Arctic because of his "contrarian nature." Textbooks put the first eyewitness stories of Arctic air pollution in the 1950s, but Garrett was convinced that there must have been earlier accounts. He was right. "Our instinctive reaction is to believe the world was a cleaner place a hundred and thirty years ago. But industry was already darkening the snow and skies of the far North," he says.

He has seen modern Arctic hazes for himself. His first experience was on a trip to the Arctic in 1998 on a University of Washington aircraft designed to study aerosols, the fine particles carried in the air, and clouds. "It was utterly new to me and completely surprising," he says. "The plane would descend and the horizontal view would go from being absolutely crystal clear visibility, out to infinity, to looking like downtown Los Angeles." The air pollution comes in distinct layers. "It is very stark," says Garret.

I had been led to Timothy Garrett by his contrarian piece of historical research, but quickly progressed to a paper he had published in *Nature*.[5] It showed that when air pollution reaches the Arctic and mixes with low clouds it quickly forms a Los Angeles–style smog that can trap heat, especially in winter and, like a thick blanket, keep the surface below much warmer than it would normally be. His work led me on again to many more papers showing that the air pollution carried to the Arctic could warm it in several ways, as a cloud blanket as he had described, as a heat-absorbing dark haze in the air above the ice, and by landing on the surface of the ice and darkening it so that it warmed and melted faster.

From there it was a short step to an exciting question: if we could cut Arctic air pollution, especially fine "black carbon," an aerosol that is particularly good at soaking up the sun's rays, could we do something to slow Arctic warming? It turned out to be a hot topic. Garrett had been to a workshop a few months earlier.[6] "There is an idea that aerosols are contributing to the warming of the Arctic, so the logical conclusion is that reducing aerosol transport to the Arctic could help ameliorate Arctic warming in a way that is easier to address than carbon dioxide emissions," he explained.

Of course, I should have known that the idea that soot might help melt enormous areas of Arctic ice is not entirely new. When I was chatting with Olav Orheim, head of the Norwegian International Polar Year program, he told me that the Soviets had thought hard about it long ago. They even considered dropping soot on the ice in the spring in order to keep open an Arctic sea route across the top of Siberia. The soot would help melt the ice, making a fleet of icebreakers unnecessary. "Once they had made the calculation, they discovered that it would take over one thousand planes every day to do it, so they dropped the idea," said Orheim. "But it was actually quite seriously proposed."

Charles Zender, at the University of California at Irvine, has been studying the impact of "dirty snow" on climate for a decade. Zender, working with Mark Flanner from the University of Michigan, estimates that black carbon falling on snow and ice has contributed 0.5°C to the overall warming of between 1°C and 2°C of the Arctic since the "preindustrial era."[7] That is quite staggering. It implies that between a half and a quarter of the Arctic warming is caused by air pollution, not greenhouse gases. His colleague Mark Flanner has published data showing that the air pollution from burning fossil fuels combined with burning "biofuels" (the wood burned in stoves, mainly in the developing world) causes almost as much springtime snow-cover melt across Eurasia as does the greenhouse warming caused by carbon dioxide.[8]

The possibility of cutting air pollution to quickly cool the climate and help the ice is gaining credence, says Zender. Black carbon leaves the atmosphere in weeks, not the hundred or so years of carbon dioxide. In late 2007, he and several other scientists testified before the U.S. Congress about the possibility. "[Congressional] staffers and members of Congress who are considering legislation to mitigate climate change know that greenhouse gas reductions are necessary, and in the end, they are the only means of dealing with climate change," Zender said. "But they are aware that reducing greenhouse gas emissions has no instantaneous effect on climate. They are looking for something that can have a more immediate effect."

In his testimony, Zender explained that 30 percent of the current warming would be removed if all the different aerosols emitted by burning fossil fuels and biofuels were cut. "We must consider the combination as a whole," explains Zender, "because they are emitted by the same combustion processes; you can't stop one without the other. We have shown that if you stop emitting them, you can get a fairly quick climate benefit."

The sooner action is taken the better. "If you wait twenty years and say, OK, it really is getting warm, let's do something; then you won't get as much benefit from reducing black carbon simply because snow has melted away in the meantime. The amount of mitigation is really proportional to the area of snow and ice there is left," Zender says.

If it is all so logical, why isn't the world rushing to tackle black carbon emissions? One reason is that the calculations are complicated. Remember that volcanic eruptions can cool the earth by injecting aerosols high into the atmosphere where they act like a giant sun shade. Yet dust that falls on ice warms the earth. Depending on the type of aerosol and where it goes, aerosols may cool the planet or warm it. That is why Zender insisted that he had to look at all the aerosols produced by burning fuels. He had to take into account both effects to be sure that cutting those aerosols would cool the planet.

That cooling is hard to calculate precisely. The impact on ice cover is even harder to determine. I asked Patricia Quinn from the Pacific Marine Environmental Laboratory in Seattle, who has been making measurements of black carbon up in Barrow, Alaska, for almost a decade,[9] whether she thought we could predict the impact of cutting black carbon on the ice. "It would be hard to put a number on that," she explained, "because there is so much that affects the ice including all that natural variability, but it is something we can do in the near term. Doing something about carbon dioxide is going to take decades and decades to see an effect."

At that point I left the scientists and went in search of activists to see if there was any hope that the world would act really quickly and take a gamble on cutting Arctic air pollution. Before long I found Clean Air–Cool Planet, a pressure group based in Washington, D.C., that tackles exactly that issue and was set up by Rafe Pomerance, a well-known campaigner whose career has spanned Friends of the Earth and the U.S. State Department. He directed me to Pam Pearson, a former U.S. diplomat with a long track record in international environmental affairs. She was pushing the idea of cutting black carbon emissions at the UNFCCC climate negotiations in Poznan, Poland, in December 2008 and to the Arctic Council nations.[10]

I quickly learned that the lack of certainty makes it tougher to overcome the key obstacle to progress: the need for international action.[11] "The issue is that you can't right now say exactly how much cutting black carbon is going to get you in the Arctic," Pearson told me when I called her in Sweden where she now lives. "We can say that it will help, but of course policy makers want a cost-benefit analysis and we don't have those kinds of figures right now."

Still, she is optimistic. There is one potential win in her sights, being helped along by some startling imagery from the scientists' town of Ny Alesund back in Svalbard. When I visited Ny Alesund I experienced the pure, unspoiled Arctic the way it should be. Sitting by the old mooring mast for the zeppelin that carried the explorers Roald Amundsen and

Umberto Nobile to the North Pole in 1926, I could see right across the fjord in front of me to a huge glacier and a striking pair of matching pyramidal mountains in the far distance.

Researchers at the air pollution monitoring station on the mountain behind the tower, appropriately named Zeppelin and reached by a ski lift from the town, can tell you horror stories of what happens when pollution arrives. Then, the mountains across the fjord fade into the haze and the air turns a deep orange-brown. Soon the concentrations of pollutants are as high here in the middle of the Arctic as they would be in a big-city rush hour.

Andreas Stohl of the Norwegian Institute for Air Research in Kjeller, Norway, tracked the worst pollution event on record, in April and May 2006, by examining satellite pictures.[12] As he puts together a series of images, he can follow a great plume of smoke that blots out the Arctic land back to where it came from: enormous agricultural fires in Russia, Ukraine, Belorussia, and the Baltic nations. Fields were being burned before planting to control insects and disease.

Pearson sent me maps of agricultural fires around the world taken by heat-detecting satellites. In early spring there is just a sprinkling of fires around the northern hemisphere. Come April, they have grown into a terrifying band of fire between the 50th and 60th parallels, right across Europe and deep into Asia. The fires reach across Russia, Ukraine, and farther east. In May, the band of fire is wider and brighter still, with fires now reaching across the top of Mongolia and right to Russia's Pacific coast. Smaller patches light up the central United States and Canada, from Saskatchewan and Manitoba down through the Dakotas and on to Iowa.

This worldwide spring burn of fields comes at a bad time for the Arctic—and the world. Smoke makes its way to the Arctic and on to the ice just as the sun returns to the north. The changed albedo speeds the spring melt and kick-starts a wider feedback that melts even more ice. Ironically, as the Arctic grows warmer, it becomes more vulnerable to pollution from the south. The dome of cold air over the Arctic effectively blocks warmer air from coming up from the south in spring and summer. With a warmer Arctic, the barrier is lower and pollution from warmer areas ever farther south can get though to the Arctic. "In 2006 there was an abnormally warm springtime in the Arctic so pollution that may not have been able to penetrate the Polar Dome if it had been a colder year in the Arctic was able to get through," explains Quinn.

Black carbon from these fires is reaching Alaska, too. In April 2008, a NOAA aircraft picked up samples of haze from the northern Alaskan skies and traced them back to forest fires around Lake Baikal in Russia and field burning in Kazakhstan. Carsten Warneke and colleagues from the NOAA Earth System Research Laboratory in Boulder, Colorado, warned of yet another feedback loop. As the Arctic warms, snow vanishes earlier, the forest fire season begins earlier and lasts longer, and yet more black carbon is sent north, warming the Arctic even more.[13]

The good news is that thanks to horror stories like that 2006 pollution event, the wider political world is beginning to wake up to the climate threat from black carbon. In early April 2009, U.S. Secretary of State Hillary Clinton pledged new initiatives to hit black carbon. Later in the month the Arctic Council, at its biannual ministerial meeting, recognized the potential for slowing ice melt quickly. That is important as action needs to come from Russia, one of the council members. It won't be easy. As a report from the campaigning group Clean Air Task Force points out, the top emitters of Arctic black carbon from agricultural fires are Russia, Kazakhstan, China, the United States, and Canada. In the first three big polluting nations, open field burning is already banned but the ban is very hard to enforce.[14]

The next target is a lot harder: the fine black carbon that pours out of diesel engines in countries all around the world, including the United States, which is responsible for about 6 percent of the world total. The dirtiest forms of diesel are being slowly phased out and particle traps that remove black carbon particles before they leave the engine exhaust are being mandated. The problem is that new regulations are going to be too late for the Arctic. "You need to accelerate it and you need to retrofit. Retrofitting is the big issue," says Pearson. Some of the Arctic nations are showing willingness to move forward and others are not. Bringing public pressure to bear will be critical. So too will the future message from the sea ice. If more spectacular collapses in the ice area seize the public imagination and the attention of politicians, then action might be fast.

"The Arctic nations are becoming real leaders on this question because the Arctic is coming apart," says Pomerance.

My encounter with Baron Nordenskiöld has led me far and left me optimistic. Something can be done. We don't know how big the impact will be, but it might slow warming and it might slow the ice melt. Every extra bit of time is time to find better solutions. I've also grown fond of the baron. Reading his travel journals I can't help feeling envious of how innocently he went on his adventures to unexplored places in a great wild world, where he never dreamed that "ice dust" could have come all the way from Europe's industrial furnaces. He describes setting off on the Greenland ice with his faithful servant, "the Lapp Lars," who carried "a wooden club, with which he had slain more than twenty-five brown bears, full of marks from their teeth; and his eyes sparkled at the thought of encountering a white one."[15] We would spare the white bears now, but with the same spirit we might yet be able to keep the black carbon away from the white ice.

Chapter Seventeen

THE FUTURE OF THE ARCTIC

When I began this book, not long after that first fateful encounter with a polar bear, I saw a sped-up movie of the Arctic and its ice. I remember my surprise at how the ice whirled around the Arctic and flew through the channels between the Canadian islands as a year ran by in a few minutes.

That movie was shown by Douglas Bancroft, the enthusiastic new director of the Canadian Ice Service at a conference in Washington, D.C., in August 2007. He had put it together before anyone knew that the ice would just go on shrinking away that summer. Bancroft did have another catastrophe to report: a nine-mile-long chunk of the Ayles Ice Sheet, which had been sitting there for 3,000 years, had recently broken away from the north of Ellesmere Island.

As I approached the end of this book, I called Bancroft to see what he had made of the eighteen months since that conference. He was still enthusiastic. "Ice is hot," he said. "I've never been busier and I've never been less bored in my whole life." I mentioned that enormous ice sheet breaking away.

"Yes, we did lose Ayles over two years ago," he said. "But last year [2008] we lost three ice sheets. Losing one is interesting, but losing three in a single summer! A whole multiyear ice shelf pulled away from the Arctic archipelago. We have a billion-ton iceberg that calved off the Petermann glacier which is now down at 64° north. We had Beaufort Sea multiyear ice fragmenting and blowing away. We had holes in the ocean open up in the fairly far north." I could almost see the movie of the Arctic falling apart. But what of the future?

"I was stunned by 2007," Bancroft said. "The summer of 2008 was very close. I still have great difficulty to forecast what's going to happen on an annual and interannual basis. But it is very clear that the projections that were made five years ago that the Arctic will be ice-free in 2100 are wrong. It's going to happen a lot sooner than that. Some forecasts talk about 2015 as being seasonally ice-free and others 2040 or 2045. I don't know precisely where it will lie between these two bookends. But after seeing another year I may have an opinion."

That summary of the central issue of this book fits everything I have learned. The Arctic ice is going, but we don't know quite how fast it will go. Predicting what will happen in any one year remains tough. With new data and a little more time to see how the natural variability of the Arctic interacts with the underlying thinning, the future may be a lot clearer. One more year is important. The winter of 2009 was very cold and built ice, but by August the area of surviving ice was below that of 2008. What happens next may tell us whether we are on a trajectory to ice-free summers in 2015 or somewhere closer to 2045.

Whichever route the ice is headed on, it is going too fast for all the creatures that depend on it. There are the iconic big beasts of the Arctic: the polar bear, walrus, narwhal, ringed seal, and the beluga and bowhead whale. And there are the smaller creatures we scarcely know, but which larger life relies on, from the Arctic cod down to the cold-water copepods and the ice algae at the base of the Arctic food chain. All will have to struggle to adapt and survive in the coming transformation of the Arctic. A transformation is what it will be—if you are able to take a philosophical view of the future. Louis Fortier, who had explained the mysteries of Arctic cod to me, put it like this:

"If you look at it simply from the point of view of biological productivity, that will increase as the ice disappears. It's just that the life there, the specialists which we are all fond of, like the polar bear, walrus, and some other species which we all have in our unconscious mind, are going to get into trouble. For the rest, the ecosystem is going to be much more productive."

He is right, of course, but is there comfort in knowing that the polar bear will be replaced by the killer whale as the Arctic's top predator, in a sea alive with southern fish? I am not so sure. But less ice will most likely mean sunnier, more productive oceans. There are already seasonally ice-covered seas farther south, in the Gulf of St. Lawrence, for example, which are very different from the Arctic now but which it may come to resemble in the future.

For the people of the Arctic the opening seas may also bring a transformation, but of what kind is even more uncertain than the fate of the polar bear, as it is tied up with the global economy and regional and international politics. I do not make too much of the possibility of military conflict or fights over the riches of the North Pole. Most of the wealth of the Arctic lies within clear boundaries already, and there are rules laid out in the Law of the Sea for those who want to claim more.

The oil, gas, mining, and shipping industries may all benefit from the thinning ice as it makes the Arctic more accessible. But I'm not going to predict an automatic bonanza for the Arctic people. I have just seen too many ghost towns and too many of the perils of relying on resource development.

Pyramiden, the mine in the sky, is the greatest ghost town of all. It is not easy to get to: you must take a small boat in Svalbard for a three-hour journey across the Isfjord.

Pyramiden lies at the foot of a vast pyramid-shaped mountain, topped off by a giant square mass of cliffs that recall an Aztec temple.[1] You can spot the mine long before the boat reaches the shore. Its entrance is halfway up the mountain, where two covered conveyor systems snake steeply up the slope to meet a great black heap of spoil running down over the red-brown rock. The coal seams are buried inside the mountain rather than down under the ground. Close to the water are huge gantry cranes for loading the coal and a giant power station with two tall metal chimneys. Nearby stands a once white, four-story building, its empty windows looking out to the sea over heaps of rusted cargo containers. A vertical line of giant Russian letters spells out "PYRAMIDEN" on its front.

This was where the Soviet Union used everything in its power to subdue the Arctic and lost. No one lives in Pyramiden now, but it was once home to an army of 800 proud Soviet miners and their families. Thousands of tons of soil were brought in from Murmansk so that grass could be grown. In summer the silent fields are still lush with purple-stemmed grasses swaying in the wind, but the pigs they once fed are long gone. You pass the rows of heated greenhouses where vegetables were grown. Deeper inside the town are two heated swimming pools, schools, and rows of tall apartment blocks and administration buildings. The giant power station had plenty of coal to keep everyone warm

Along the corridors of the cultural hall, the Soviet Union has not yet ended. Black-and-white photographs line its walls, some with explanatory titles: "Visit of Norway Ambassador," says one, beneath a picture of a man in heavy glasses standing at a podium. There are photographs of people playing chess, doing gymnastics, parading with flags, and folk dancing. There are books on the library shelves and one of the card indexes still stands open as though the librarian had slipped away for a moment. The library closed in 1991. A piano, now slightly out of tune, stands in the music room; an old gramophone record from the "Melodiya" label lies in a corner covered in dust.

Our guide is one of a small party of Russian historians who spend the summer here and dream that Pyramiden may be reborn as a tourist destination. He carries an old rifle slung across his back, muzzle down, in case we run into a bear, and addresses our small group of visitors as "dear friends." He explains that one of the big apartment blocks was for single women and nicknamed Paris, one for single men was called London. Another, the Crazy House, was for families with young children. There are no children here now, but the building is still noisy, alive with the sound of kittiwakes that have built their nests on the high window ledges.

At the heart of Pyramiden is a large stone bust of Lenin. In front of him lies a field of Soviet grass, then a long wall of tall apartment buildings above which hover the glaciers and snow-covered mountains of Spitsbergen as though they were a mural painted in the sky.

Despite everything the Soviet Union did, the cold reality of the Arctic intruded. It proved impossible to maintain the town once the new Russia arrived with "market forces." The miners had been well paid, much respected, and totally cared for by the State. One day in 1991 the last of them were told that they had to leave very quickly. The town was abandoned. "I read some of their letters," says our guide. "They were asking,

'What is to become of us, where are we to go?' It brought tears to my eyes."

Greenland has a mining ghost town, too, at Qullissat. Hundreds of Inuit were encouraged to leave their tiny communities and move to the coal mines here. By the mid-1960s the town had 1,400 inhabitants, an urban lifestyle, and a big store selling imported goods. But the mine proved too expensive. In 1966 the decision was made to close it down and move all the people out. No one lives here now, but it is a beautiful spot. You can rent one of the old houses and sit here in the silence, watching the icebergs sail by.

There is a sadness to these empty places that I felt again when I talked to Charlie Lyall, president and CEO of the economic development arm of the Inuit Association of Kitikmeot, the most westerly region of Nunavut. I called him in Taloyoak, one of its small communities, as I had heard that Nunavut's first diamond mine, opened in 2005 with a commitment from Tiffany to buy its diamonds and a ribbon cutting by Prime Minister Stephen Harper, had abruptly closed down. The mine had been a bright hope for Nunavut's development as diamond mines a little farther south had been very successful, but it had not been as productive as hoped. Lyall explained that people from all six communities in the region had been working at the mine. It had meant jobs and something more. "When Inuit are making a meaningful living, it's a lot better," Lyall explained. "You see the community being much more vibrant, everybody feels much better about themselves, and life is good."

Lyall does not let the closure get him down. "I'm optimistic about the future for this region," he says. "In the long term, mining is going to be the way to get the economy to grow here. The potential is also there for oil and gas." There are gold, copper, zinc, and silver mines being planned in the Kitikmeot area that have the potential to bring local jobs. The really big chance for an economic transformation for Nunavut and Greenland remains with oil and gas, though, a transformation that Alaska has already made. Environmentalists may not much like to hear that, but development and protection of the environment have to go together if the people of the north are to prosper.

Whether oil or gas development goes ahead, and how quickly, depends on the cost of developing new, safe Arctic technology, the cost of building, installing, and maintaining the technology, the size of the reservoir (only a giant long-term payoff can justify a big capital cost), the cost of taking the oil to market, and the degree of political support and stability. And, of course, the price of oil.

Although it would be good to think that the oil industry will bring long-term wealth to the Arctic, my bet is that the oil and gas boom will be short-lived and will not go far beyond the shallow seas of Russia and perhaps some of the regions close to the Alaskan shores. The reason is that the Arctic's message is finally getting through to the wider world. The ice is melting away and it is shouting, "Your planet is in danger." A price is going to be put on carbon emissions and that will drive the search for alternatives to oil. My bet is that changing energy use will make all but the easiest to find Arctic oil too expensive in the next few decades. I'm not predicting the end of oil at all, just the end of very expensive oil from extreme environments.

Skeptics will laugh. They will point to the growing demand from China and India and the shortage of oil and all the extrapolations that show oil demand and prices will grow far into the future. For the skeptics, the Forest of Dean is a good place to go. It is an old oak forest on the border of Wales which would not be there if one of Britain's greatest heroes hadn't also been stuck in believing the future would be more of the present.

In 1802, Admiral Lord Nelson toured the Forest of Dean looking for timber to refit his flagship, HMS *Victory*. He needed wood for new battleships, too, and each required 6,000 mature oak trees to build. The forest was in an appalling state. But he managed to find the timber he needed and went on to defeat the combined fleet of France and Spain at the Battle of Trafalgar in 1805. Nelson died during the battle, but his concerns about the shortage of old oak weren't forgotten. Thirty million acorns were planted across 11,000 acres of land to provide timber for the battleships that the Royal Navy thought they might need in fifty to a hundred years time. Some oaks were planted close together so they would grow straight and could be cut for planking; others so that their branches would curve upward and be right for a battleship's massive ribs. They are still there. Iron and steel replaced wood, transformed the world, and made Nelson's oaks redundant. The only wood taken from the forest was to repair HMS *Victory*, now in a dockside museum.

What might come along and make Arctic oil as archaic as wooden warships? There are many possibilities. (I should disclose that I have another job helping run a small U.S. company that brings together investors and innovators.)[2] One that's being talked about a lot is "renewable gasoline," that is, using genetically engineered algae to grow "green oil" that could go straight into existing refineries. Then there are electric automobiles, where China has surprised everyone by pushing to take a lead. Given the size of the prize for a successful invention, the worldwide talent, and growing pressure to cut carbon emissions, it is a reasonable bet that new technologies will appear. Oil will play a major part in a cocktail of new ways to provide energy, but very expensive oil will no longer make sense a decade or two from now. Every time the oil price goes up and the Arctic looks more attractive, alternative energy sources will also look more attractive. Oil executives won't be unhappy if something new comes along. One said to me, "There is a saying in our industry, 'the Stone Age didn't end because the world was running out of stone.' When we find something better, we'll invest billions."

Any Arctic oil boom may last only a few decades, although gas and shipping may run longer. Patricia Cochran, head of the Inuit Circumpolar Council, has three bullet points of wise advice from Inupiat experience of the oil industry on Alaska's North Slope. It reads:

- DO NOT ABANDON YOUR PREVIOUS LIVES.
- MAINTAIN TRADITIONAL WAYS OF LIFE.
- EXTRACT LOCAL BENEFITS WHILE OIL IS STILL FLOWING.

Oil, gas, and mining can provide a huge lift to a community, but the community that thinks they will be there forever, or even worse, lets outsiders take all the benefits, is in for trouble. Of course, it could turn out that the Arctic will be the world's next great energy province and that after its oil and gas run out, the open seas in a warmer planet will ease the mining of methane hydrates from the seabed all the way to the North Pole. It is not impossible, but it is a future that represents a complete failure of the imagination—one where we cannot find better ways to work together to conserve and create energy.

What else might bring the Arctic, with its fast-growing population of young people, a good sustainable life? "Everybody talks tourism, but it is expensive to travel up here. So tourism is not that great," says Lyall. "There is always the hunting and trapping, but there again with everything that's going on with the animal rights groups there is not a huge future in it either. Inuit will continue to hunt and fish for their own sustenance, but they can't make a living out of it." Lyall is alluding to that bitter clash between southerners and hunters over the killing of seals and hunting for fur. "You probably have a cockroach protection society in London," he adds, in a wry tone.

Fishing is a very important part of the Greenland economy, and it is likely to grow further as seas warm.[3] It could be bigger across the other side of Baffin Bay, too, but that requires investment in harbors, which is only just beginning. "We are still only getting a small percentage of the fish quota from our waters," says John Amagoalik, the Inuit leader from Iqaluit. "A lot of southern Canadian interests have been given fish allot-ments in Baffin Bay and our surrounding waters. We are hoping that the government of Canada will start giving us more of those quotas." Inuit art, too, will "always be a part of our economy."

Several Arctic commentators have looked ahead and laid out sce-narios for how the future may develop, most notably Lawson Brigham, a former U.S. icebreaker captain and lead author of the Arctic Marine Shipping Assessment, and Bjørn Brunstad at the Oslo-based consultancy Econ.[4] Putting aside dark scenarios of world conflict, all reach the same broad conclusions.

The key issue is whether or not there will be a high value placed on Arctic oil, gas, and mineral resources, along with easier access to them by sea. If that comes true, then the question becomes whether or not the people who live in the Arctic will benefit from the wealth that surrounds them and can ensure that their environment is protected as development proceeds. One route leads to an Arctic that is a part of the global com-munity, with living standards to match. The other leaves Arctic people living among transient workers from the south in a ruined landscape. If the world places a low value on Arctic resources, then the people of the Arctic will have to find new ways to live from the wildlife of their rapidly changing environment, or face a tough life on welfare. We will see either a sustainable Arctic or an abandoned Arctic.

The greatest issue of all is time. We don't know how fast the changes in the Arctic are coming. They may come too fast to allow people and animals to adapt and too fast for governments to react and protect what remains. The worst of all possible worlds is an abandoned, ruined Arctic. If we are lucky, the Arctic may give us more time. If the world could then work together, vigorously cutting air pollution that warms the Arctic, and if that buys more time to follow up with aggressive actions on greenhouse gases, then some of the summer ice might yet be saved. But with every year of delay, we need ever stronger action, and the chances of success grow less.

One day in Tasiilaq, Greenland, I sat and chatted with a few teenagers from the local soccer teams (I had been watching their games; they had both a boys team and a girls team and a crowd of vociferous supporters). I asked what would happen as the weather warmed and the ice in the bay melted away. "We don't mind, we'll find oil and we'll be as rich as Saudi Arabia," I was told. "What will you do then?" I asked. "I'll buy a boat and go hunting with my Granddad," said one. "I'll have my own helicopter," said a second, who was already planning to train as a pilot. "I'll go shopping in New York," added the one girl who had joined us, who I'd earlier seen score a particularly lethal goal. They left me with a bouquet of white Arctic cotton grass that their smaller brothers and sisters had gathered.

I hope their dreams come true. They'll have to navigate a tricky route to a new sustainable future if that wealth does not flow quite as freely as they hope. Looking beyond their soccer field ground to the harbor at Tasiilaq and its little store selling "country foods," I don't think that narwhal and polar bear will be on sale there much longer. Great change is coming.

SOURCES

Interviews with scientists, engineers, and policy makers provided the key source material for this book. Over one hundred people kindly agreed to be interviewed, answer my questions, or help me with research material (approximately 40 percent of them from the United States, 20 percent from Canada, and 25 percent from Norway, Denmark-Greenland, Finland, and Russia). Many of them are quoted in the text. Their names are all given below to register my profound gratitude (my only regret is that I cannot include the names of officials from the Foreign Offices or State Departments of four nations who provided off-the-record briefings on sensitive border or military issues). Any errors anywhere in this book are, of course, entirely my own.

In contacting these many individuals it was always very helpful that this book was chosen to be a part of the International Polar Year (IPY) Polar Books project of the United Nations Environment Program (see www.unep.org/publications/polarbooks).

John Amagoalik, Qikiqtani Inuit Association, Iqaluit, Nunavut, Canada; **Steven Amstrup**, U.S. Geological Survey Alaska Science Center, Anchorage, Alaska, United States; **Elena Andreeva**, Institute for System Analysis of the Russian Academy of Sciences, Moscow, Russia; **Kevin Arrigo**, Stanford University, Stanford, California, United States; **Carin Ashjian**, Woods Hole Oceanographic Institution, Woods Hole, Massachusetts, United States; **Natalie Asselin**, University of Manitoba, Winnipeg, Manitoba, Canada; **Douglas Bancroft**, Canadian Ice Service, Ottawa, Ontario, Canada; **John Baglow**, Ottawa, Canada; **David Barber**, University of Manitoba, Winnipeg, Manitoba, Canada; **Søren Basbøll**, Tasiilaq, Greenland; **Benoît Beauchamp**, The Arctic Institute of North America, Calgary, Alberta, Canada; **David Beerling**, University of Sheffield, Sheffield, United Kingdom; **David Benton**, Marine Conservation Alliance, Juneau, Alaska, United States; **Jørgen Berge**, The University Centre in Svalbard, Longyearbyen, Norway; **Bjørn Brunstad**, Econ, Oslo, Norway; **Ian Buist**, SL Ross Environmental Research, Ottawa, Ontario, Canada; **Patricia Cochran**, Chair, Inuit Circumpolar Council, Anchorage, Alaska, United States; **Ruth Curry**, Woods Hole Oceanographic Institution, Woods Hole, Massachusetts, United States; **Jan Inge Dalane**, StatoilHydro, Stavanger, Norway; **Jack Dibb**, University of New Hampshire, Durham, New Hampshire, United States; **George Divoky**, Friends of Cooper Island, Seattle, Washington, United States; **George Edwardson**, President, Inupiat Community of the Arctic Slope, Barrow, Alaska, United States; **Steven Ferguson**, University of Manitoba, Winnipeg, Manitoba, Canada; **Hans Christian Florian**, Tasiilaq Hospital, Greenland; **Bruce Forbes**, University of Lapland, Rovaniemi, Finland; **James Ford**, McGill University, Montreal, Quebec, Canada; **Louis Fortier**, Université Laval, Quebec, Quebec, Canada; **Tim Garrett**, University of Utah, Salt Lake City, Utah, United States; **Jean Claude Gascard**, Laboratoire d'Océanographie et du Climat, Paris, France; **Rudiger Gerdes**, Alfred Wegener Institute, Bremerhaven, Germany; **Katherine Giles**, University College London, London, United Kingdom; **Rolf Gradinger**, University of Alaska, Fairbanks, Alaska, United States; **Patty Gray**, National University of Ireland, Maynooth, Kildare, Ireland; **Ove Gudmestad**, Norwegian University of Science and Technology, Trondheim, Norway; **Christian Haas**, University of Alberta, Edmonton, Alberta, Canada; **Otto Habeck**, Max Planck Institute for Social Anthropology, Halle/Saale, Germany; **Edward Hanna**, University of Sheffield, Sheffield, United Kingdom; **Ailsa Henderson**, University of Toronto at Mississagua, Ontario, Canada; **Rod Hobbs**, National Marine Mammal Laboratory, Seattle, Washington, United States; **Ian Howat**, Ohio State University, Columbus, Ohio, United States; **Toke Høye**, University of Aarhus, Aarhus, Denmark; **Rob Huebert**, University of Calgary, Calgary, Alberta, Canada; **Mark Jacobson**, Stanford University, Stanford, California, United States; **Chadwick Jay**, U.S. Geological Survey Alaska Science Center, Anchorage, Alaska, United States; **Martin Jeffries**, National Science Foundation, Washington, D.C., United States; **Philip Jones**, Climatic Research Unit, University of East Anglia, Norwich, United Kingdom; **Nancy Kinner**, University of New Hampshire, Durham, New Hampshire, United States; **Kit Kovacs**, Norwegian Polar Institute, Tromsø, Norway; **Rick Krishfield**, Woods Hole Oceanographic Institution, Woods Hole, Massachusetts, United States; **David Lawrence**, National Center for Atmospheric Research, Boulder, Colorado, United States; **Seymour Laxon**, University College London, London, United Kingdom; **Jonathan Lindsay**, J P Kenny Caledonia Ltd., Aberdeen, United Kingdom; **Ron Lindsay**, University of Washington, Seattle, Washington, United States; **Brian Lumsden**, Stena Drilling, Aberdeen, United Kingdom; **Charlie Lyall**, Kitikmeot Cooperation, Cambridge Bay, Nunavut, Canada; **Ron Macnab**, Canadian Polar Commission, Dartmouth, Nova Scotia, Canada; **James Maslanik**, University of Colorado at Boulder, Boulder, Colorado, United States; **Wieslaw Maslowski**, Naval Postgraduate School, Monterey, California, United States; **Svein Mathiesen**, International Centre for Reindeer Husbandry, Kautokeino, Norway; **Larry Mayer**, University of New Hampshire, Durham, New Hampshire, United States; **Louis McComber**, Serpentine Communications, Sutton, Quebec, Canada; **Gordon McCreary**, Baffinland, Toronto, Ontario, Canada; **Jacqueline McGlade**, European Environment Agency, Copenhagen, Denmark; **Amy Merten**, Office of Response and Restoration, NOAA, Seattle, Washington, United States; **Dines Mikaelsen**, hunter, Tasiilaq, Greenland; **CJ Mundy**, University of Quebec at Rimouski, Rimouski, Quebec, Canada; **Tavi Murray**, University of Wales Swansea, Swansea, United Kingdom; **Rebecca Noblin**, Center for Biological Diversity, Anchorage, Alaska, United States; **Erik Olsen**, Institute of Marine Research, Bergen, Norway; **Olav Orheim**, Executive Secretary, Research Council of Norway, Oslo, Norway; **Anders Oskal**, International Centre for Reindeer Husbandry, Kautokeino, Norway; **Michael Paulin**, IMV Projects Atlantic, St. John's, Newfoundland, Canada; **Pam Pearson**, Clean Air–Cool Planet, Sweden; **Tad Pfeffer**, University of Colorado at Boulder, Boulder, Colorado, United States; **Rafe Pomerance**, Clean Air–Cool Planet, Climate Policy Center, Washington, D.C., United States; **Patricia Quinn**, Pacific Marine Environmental Laboratory, NOAA, Seattle, Washington, United States; **Jackie Richter-Menge**,

Cold Regions Research and Engineering Laboratory, Hanover, New Hampshire, United States; **Eric Rignot**, NASA Jet Propulsion Laboratory, Pasadena, California, United States; **Vladimir Romanovsky**, University of Alaska, Fairbanks, Alaska, United States; **Peter Roopnarine**, California Academy of Science, San Francisco, California, United States; **Elisabeth Sæther**, Bellona, Oslo, Norway; **Gunnar Sander**, Norwegian Polar Institute, Tromsø, Norway; **Tatiana Saksina**, World Wildlife Fund, Oslo, Norway; **Igor Semiletov**, Pacific Oceanological Institute, Russian Academy of Science, Russia; **Mark Serreze**, National Snow and Ice Data Center, Boulder, Colorado, United States; **Whit Sheard**, Pacific Environment, Anchorage, Alaska, United States; **Andrew Shepherd**, University of Edinburgh, Edinburgh, United Kingdom; **Koji Shimada**, Japan Agency for Marine-Earth Science and Technology, Yokosuka, Japan; **Roald Byhre Sirevaag**, StatoilHydro, Stavanger, Norway; **Mary Simon**, President, Inuit Tapiriit Kanatami, Ottawa, Ontario, Canada; **David Smith**, University of Oxford, Oxford, United Kingdom; **Mark Smith**, Defence Academy of the United Kingdom, Shrivenham, Wiltshire, United Kingdom; **Janne Søreide**, The University Centre in Svalbard, Longyearbyen, Norway; **Florian Stammler**, University of Lapland, Rovaniemi, Finland; **Michael Steele**, University of Washington, Seattle, Washington, United States; **Anders Stenbakken**, SIKU Extreme Arctic Challenge, Tasiilaq, Greenland; **Fiamma Straneo**, Woods Hole Oceanographic Institution, Woods Hole, Massachusetts, United States; **Julienne Stroeve**, National Snow and Ice Data Center, Boulder, Colorado, United States; **Reko-Antti Suojanen**, Aker Arctic Technology Inc., Helsinki, Finland; **Ståle Sveinungsen**, Vardø Vessel Traffic Services, Vardø, Norway; **Marco Tedesco**, City College of New York, New York, United States; **Robert Thompson**, REDOIL, Alaska, United States; **Mary Louise Timmermans**, Woods Hole Oceanographic Institution, Woods Hole, Massachusetts, United States; **Annis May Timpson**, University of Edinburgh, Edinburgh, United Kingdom; **John Toole**, Woods Hole Oceanographic Institution, Woods Hole, Massachusetts, United States; **Mead Treadwell**, U.S. Arctic Research Commission, Anchorage, Alaska, United States; **Robert D. Tustin**, Lloyd's Register Asia, Busan, South Korea; **Virginie Vaté**, University of Paris, Paris, France; **Peter Wadhams**, University of Cambridge, Cambridge, United Kingdom;

Katey Walter, University of Alaska, Fairbanks, Alaska, United States; **Muyin Wang**, University of Washington, Seattle, Washington, United States; **Margaret Williams**, World Wildlife Fund, Anchorage, Alaska, United States; **Charlotte Wisnes**, The North Atlantic Marine Mammal Commission, Tromsø, Norway; **Shaye Wolf**, Center for Biological Diversity, San Francisco, California, United States; **Charles Zender**, University of California, Irvine, California, United States; **Nikita Zimov**, Northeast Science Station, Cherskii, Russia.

Further information came from some 250 public lectures I attended at conferences over an eighteen-month period, including: "Impact of an Ice-Diminishing Arctic on Naval and Maritime Operations" (2007), Washington, D.C., United States; "The Arctic Energy Summit Technology Conference" (2007), Anchorage, Alaska, United States; "Arctic Frontiers" (2008), Tromsø, Norway; "The Inhabited Arctic: Humanities and Social Science Research in the Circumpolar North" (2008), London, United Kingdom; "Earth's Climate: Past, Present and Future" (2008), London, United Kingdom; "Arctic Passion Seminar" (2008), Helsinki, Finland; "Arctic Change" (2008), Quèbec, Canada; "American Geophysical Union Fall Meeting" (2008), San Francisco, United States; "Arctic Frontiers" (2009), Tromsø, Norway.

Written materials, especially scientific papers in the journals *Geophysics Research Letters*, *Nature*, *Science*, and *Nature Geoscience*, were very valuable, along with English-language newspapers from Nunavut, Greenland, Norway, Russia, Alaska, and the Barents Area. Two e-mail newsletters from the Institute of the North in Anchorage, Alaska (*The Top of the World Telegraph* from Nancy Hemsath and *Arctic Synergy* from Kristina Baiborodova), were particularly important and provided me with many leads that I would never otherwise have found. A blog called "Arctic Economics" (http://benmuse.typepad.com/arctic_economics/) was also very valuable and led to many new sources. Other blogs with valuable viewpoints included http://www.thearctic.is/ and http://arcticblog.arcticfocus.com/.

This book benefited greatly from early reading and editing by Elizabeth Else, associate editor of *New Scientist* magazine. To her goes the credit of beginning the book with the "Bear on the Beach," the story of my first encounter with a polar bear, along with many other excellent ideas. At Smithsonian Books, executive editor Elisabeth Kallick Dyssegaard patiently worked through many drafts many times, adding enormous value and keeping me on track, assisted by the resourceful Kathryn Antony who made sure that I reached the finish line, almost on time. In London, Ed Faulkner, editorial director of Virgin Books, contributed many wise insights.

The diagrams were all created by Nigel Hawtin, graphics editor of *New Scientist* magazine.

Sources for the diagrams were as follows: Ice drift stations, Russian State Museum of the Arctic and Antarctic, St. Petersburg and Fram Museum, Oslo; Arctic sea ice, U.S. National Snow and Ice Data Center and DAMOCLES (www.seaice.dk); Russian Oil and Gas Fields, Gazprom and Lukoil; Yamal reindeer migration route, Stammler, Florian (2005). *Reindeer Nomads Meet the Market: Culture, Property, and Globalisation at the "End of the Land."* Berlin: LIT Verlag; Arctic sectorial and equidistance divisions, Ron Macnab, Canadian Polar Commission.

Anders Oskal and Svein Mathiesen of the International Centre for Reindeer Husbandry, Kautokeino, Norway, kindly gave permission to quote the words of reindeer herders from their book on the EALÁT project.

Finally, a very special thank you to the European Environment Agency for arranging a visit to southeast Greenland.

NOTES

INTRODUCTION: THE BEAR ON THE BEACH

1. Bancroft, Douglas (2007). "Impact of an Ice-Diminishing Arctic on Naval and Maritime Operations." Washington, D.C., July 10–12.
2. Lopez, Barry (1987). *Arctic Dreams*. Bantam Books. A classic, intense evocation of the Arctic.
3. For the threat to the polar deserts that lie at the edge of the permanent ice see the work of D. A. "Skip" Walker, director of the Alaska Geobotany Center at the University of Alaska in Fairbanks and the Circumpolar Arctic Vegetation Mapping Project.

CHAPTER ONE: TURN THE WORLD ON ITS SIDE

1. To learn to love mosquitoes, read this delightful book: Pielou, E. C. (1994). *A Naturalist's Guide to the Arctic*. Chicago: University of Chicago Press.

CHAPTER TWO: IN AN INUIT LAND

1. A beautifully written account of the relocation and the feelings of those who experienced it can be found in McGrath, Melanie (2006). *The Long Exile: A Tale of Inuit Betrayal and Survival in the High Arctic*. New York: Alfred A. Knopf. There is also a wonderful and deeply researched book that tells you everything about the island on which Grise Fiord is located and includes a thorough account of the relocation: Dick, Lyle (2001). *Muskox Land: Ellesmere Island in the Age of Contact*. Calgary, Alberta: University of Calgary Press.
2. Amagoalik, John (2007). *Changing the Face of Canada*. Iqaluit: Nunavut Arctic College. This biography is an important source for the creation of

Nunavut, and its extensive notes by Louis McComber provide the best timeline of the north's political development.

3. An incredible photographic record of the Alaskan whale hunt can be found in Hess, Bill (1999). *Gift of the Whale: The Inupiat Bowhead Hunt, a Sacred Tradition.* Seattle: Sasquatch Books.

4. *Survey of Living Conditions in the Arctic: Inuit, Saami, and the Indigenous Peoples of Chukotka.* Poppel, Birger, Jack Cruise, Gérard Duhaime, Larisaa Abryutina. 2007. SLiCA Results. Anchorage: Institute of Social and Economic Research, University of Alaska, Anchorage.

5. *Time* magazine, "Milestones." October 4, 1954.

6. See note 2 above.

7. A useful book that covers development up to the creation of Nunavut is. Dahl, Jens, Jack Hicks, and Peter Jull (2000). *Inuit Regain Control of Their Lands and Their Lives.* International Work Group for Indigenous Affairs.

8. Wells, Zachariah (2004). *Unsettled.* Ontario: Insomniac Press. A book of poetry that wonderfully evokes Iqaluit and other High North scenes.

9. *The Nunavut Handbook: Travelling in Canada's Arctic* (2004). Iqaluit: Ayaya Marketing & Communications.

10. Henderson, Ailsa (2007). *Nunavut: Rethinking Political Culture.* Vancouver: UBC Press. A book full of insights about the new political process.

11. Loukacheva, Natalia (2007). *The Arctic Promise: Legal and Political Autonomy of Greenland and Nunavut.* Toronto: University of Toronto Press.

12. See note 4 above.

13. Björkstén, Karin, Daniel Kripke, and Peter Bjerregaard (2009). "Accentuation of Suicides but Not Homicides with Rising Latitudes of Greenland in the Sunny Months." *BMC Psychiatry* 9:20.

14. *RCMP Review of Allegations Concerning Inuit Sled Dogs* (2006). Final Report.

15. Nunavut Tunngavik Inc. NTI press release

16. Berger, Thomas (2006). *The Nunavut Project.* Conciliator's Final Report.

17. UN Permanent Forum on Indigenous Issues: Indigenous Children's Education and Languages. (2008) *Indigenous Children's Education and Indigenous Languages* (document E/C.19/2005/7).

18. Krupnik, Igor, and Dyanna Jolly, eds. (2002). *The Earth is Faster Now: Indigenous Observations of Arctic Environmental Change.* Fairbanks, Alaska: Arctic Research Consortium of the United States; Nickels, S., C. Furgal, M. Buell, and H. Moquin (2006). *Putting the Human Face on Climate Change: Perspectives from Inuit in Canada.* Ottawa: Inuit Tapiriit Kanatami, Nasivvik Centre for Inuit Health and Changing Environments at Universitè Laval and the Ajunnginiq Centre at the National Aboriginal Health Organization.

19. The Alaska Village Erosion Technical Assistance Program from the U.S. Army Corps of Engineers provides regular risk assessments.

20. *The Mayer Report on Nunavut Devolution* (2007). For the Ministry of Indian Affairs and Northern Development.

CHAPTER THREE: NOMADS OF THE YAMAL

1. For a truly magical account of living among the Eveny reindeer herders, nothing beats this famous book: Vitebsky, Piers (2005). *Reindeer People: Living with Animals and Spirits in Siberia*. London: HarperCollins. For a valuable technical report see Jernsletten, Johnny-Leo L., and Konstantin Klokov (2002). *Sustainable Reindeer Husbandry*. Arctic Council 2000-2002. Tromsø: Centre for Sámi Studies, University of Tromsø.

2. Statistics on herding along with many atmospheric pictures of the reindeer herders and opinions expressed in their own words are found in a superb publication from the International Centre for Reindeer Husbandry, *EALÁT: Reindeer Herder's Voice: Reindeer Herding, Traditional Knowledge and Adaptation to Climate Change and Loss of Grazing Land*, edited by Oskal, Anders, Johan Mathis Turi, Svein Mathiesen, and Philip Burgess (2008). See page 5.

3. Krupnik, Igor (1993). *Arctic Adaptations: Native Whalers and Reindeer Herders of Northern Eurasia*. Hanover, N.H.: University Press of New England.

4. Stammler, Florian et al. (2008). "Humans and Reindeer on the Move." In *Nomadic Peoples*, published by Berghahn Journals.

5. A full account of the Yamal reindeer herders' year, which I have drawn on here, is found in chapter 3 of a wonderful book, Stammler, Florian (2005). *Reindeer Nomads Meet the Market: Culture, Property and Globalisation at the "End of the Land."* Münster: LIT Verlag.

6. Golovnev, Andrei V. and Gail Osherenko (1999). *Siberian Survival: The Nenets and Their Story*. Ithaca, N.Y.: Cornell University Press. The classic account of the Nenets. See page 98 for this story.

7. Quoted in note 2 above, page 53.

8. The "Declaration on Coexistence of Oil & Gas Activities and Indigenous Communities on Nenets and Other Territories in the Russian North" can be found on the Web site of the Arctic Centre at the University of Lapland www.arcticcentre.org.

9. Gray, Patty A., and Florian Stammler (2002). "Siberia Caught between Collapse and Continuity." *Max Planck Research* vol. 2, 55. The political issues facing indigenous people in the north of Russia are well described in the following three books, the first of which is notable for containing one

of the few English translations of the short story "About That for Which There Is No Name" by the Nenets writer Anna Nergaki.

Pika, Alexander, Jens Dahl, and Ingo Larsen, eds. (1996). *Anxious North: Indigenous People in Soviet and Post-Soviet Russia*. Selected documents, letters, and articles. Copenhagen: International Work Group for Indigenous Affairs; Slezkine, Yuri (1994). *Arctic Mirrors: Russia and the Small Peoples of the North*. Ithaca, N.Y.: Cornell University Press; Wessendorf, Kathrin, ed. (2005). *An Indigenous Parliament? Realities and Perspectives in Russia and the Circumpolar North*. Copenhagen: International Work Group for Indigenous Affairs.

10. See note 2 above, page 97.
11. Quoted in note 2 above, page 53.

CHAPTER FOUR: ADRIFT ON THE ICE

1. Drivenes, Einar-Arne, and Harald Dag Jølle, eds. (2006). *Into the Ice: The History of Norway and the Polar Regions*. Tromsø: Gyldendal Norsk Forlag, page 76. A brilliant book with many unique pictures, this account has the advantage of showing many different expeditions in their historical context so that the continuity between them can be seen.
2. Reported in the *Hartford Courant* newspaper on February 19, 1897.
3. See note 1 above, pages 82 and 84; Nansen, Fridtjof (2000). *Farthest North: The Incredible Three-Year Voyage to the Frozen Latitudes of the North*. London: Gerald Duckworth.
4. See note 1 above, page 81.
5. The standard English-language account of Nansen's life is Huntford, Roland (2001). *Nansen*. London: Abacus.
6. Papanin, Ivan (1939). *Life on an Ice Floe*. New York: Julian Messner. Now out of print, an account of bravery accompanied by eulogies to Stalin.
7. Althoff, William F. (2007). *Drift Station: Arctic Outposts of Superpower Science*. Dulles: Potomac Books. Strongly recommended for the account of scientists out on the ice. His description of Papanin's NP1 is outstanding. Contains much information that cannot be found anywhere else.
8. Russian State Museum of the Arctic and Antarctic, St. Petersburg.
9. Maloney, Sean M. "Canada's Arctic Sky Spies." *Canadian Military Journal* 9, no. 1.
10. A 110-pound intelligent seaglider, developed by researchers at the University of Washington, set a new record for an under-ice voyage in February 2009 when it surfaced off the coast of Greenland after traveling 450 miles under the ice in 51 days.

11. Stroeve, Julienne. "Arctic Sea Ice Shrinks as Temperatures Rise." *NSIDC Arctic Sea Ice News & Analysis* October 3, 2006.

CHAPTER FIVE: THE VIEW FROM BELOW

1. McLaren, Alfred Scott (2008). *Unknown Waters: A First-Hand Account of the Historic Under-Ice Survey of the Siberian Continental Shelf by USS* Queenfish. Tuscaloosa: University of Alabama Press. The only public account of a secret submarine mission, mapping waters close to Russia during the Cold War. Gripping.
2. Wadhams, Peter (1990) "Evidence of Thinning of the Arctic Ice Cover North of Greenland." *Nature* 345: 795.
3. Rothrock, D. A., Y. Yu, and G. A. Maykut (1999). "Thinning of the Arctic Sea-Ice Cover." *Geophysical Research Letters* 26: 3469–3472.
4. Maslanik, J. A., C. Fowler, J. Stroeve, S. Drobot, J. Zwally, D. Yi, and W. Emery (2007). "A Younger, Thinner Arctic Ice Cover: Increased Potential for Rapid Extensive Sea-Ice Loss." *Geophysical Research Letters* 34: L24501; Fowler, C., W. J. Emery, and J. Maslanik (2004). "Satellite-Derived Evolution of Arctic Sea Ice Age: October 1978 to March 2003." *Geoscience and Remote Sensing Letters* 1: 71–74.
5. Haas, Christian, Andreas Pfaffling, Stefan Hendricks, Lasse Rabenstein, Jean-Louis Etienne, and Ignatius Rigor (2008). "Reduced Ice Thickness in Arctic Transpolar Drift Favors Rapid Ice Retreat." *Geophysical Research Letters* 35: L17501.
6. Shimada, Koji, Takashi Kamoshida, Motoyo Itoh, Shigeto Nishino, Eddy Carmack, Fiona McLaughlin, Sarah Zimmermann, and Andrey Proshutinsky (2006). "Pacific Ocean Inflow: Influence on Catastrophic Reduction of Sea Ice Cover in the Arctic Ocean." *Geophysical Research Letters* 33 (8): L08605; Sumata, Hiroshi, and Koji Shimada (2007). "Northward Transport of Pacific Summer Water along the Northwind Ridge in the Western Arctic Ocean." *Journal of Oceanography* 63: 363–378.

CHAPTER SIX: THE LETHAL MIX

1. Perovich, Donald K., Jacqueline A. Richter-Menge, Kathleen F. Jones, and Bonnie Light (2008). "Sunlight, Water, and Ice: Extreme Arctic Sea Ice Melt During the Summer 2007." *Geophysical Research Letters* 35: L11501; Perovich, Donald K., Bonnie Light, Hajo Eicken, Kathleen F. Jones, Kay Runciman, and Son V. Nghiem (2007). "Increasing Solar Heating of the

Arctic Ocean and Adjacent Seas, 1979–2005: Attribution and Role in the Ice-Albedo Feedback." *Geophysical Research Letters* 34: L19505; Nghiem, S. V., I. G. Rigor, D. K. Perovich, P. Clemente-Colón, J. W. Weatherly, and G. Neumann (2007). "Rapid Reduction of Arctic Perennial Sea Ice." *Geophysical Research Letters* 34: L19504.

2. Steele, Michael, Wendy Ermold, and Jinlun Zhang (2008). "Arctic Ocean Surface Warming Trends over the Past 100 Years." *Geophysical Research Letters* 35: L02614.

3. Kay, Jennifer E., Tristan L'Ecuyer, Andrew Gettelman, Graeme Stephens, and Chris O'Dell (2008). "The Contribution of Cloud and Radiation Anomalies to the 2007 Arctic Sea Ice Extent Minimum." *Geophysical Research Letters* 35: L08503; Ogi, Masayo, Ignatius G. Rigor, Miles G. McPhee, and John M. Wallace (2008). "Summer Retreat of Arctic Sea Ice: Role of Summer Winds." *Geophysical Research Letters* 35: L24701.

4. Kwok, Ron (2008). "Summer Sea Ice Motion from the 18GHz Channel of AMSR-E and the Exchange of Sea Ice between the Pacific and Atlantic Sectors." *Geophysical Research Letters* 35: L03504.

5. Lindsay, R. W., J. Zhang, A. Schweiger, M. Steele, and H. Stern (2009). "Arctic Sea Ice Retreat in 2007 Follows Thinning Trend." *Journal of Climate* 21: 364.

6. Deser, Clara, and Haiyan Teng (2008). "Evolution of Arctic Sea Ice Concentration Trends and the Role of Atmospheric Circulation Forcing, 1979–2007." *Geophysical Research Letters* 35: L02504; Zhang, Jinlun, Ron Lindsay, Mike Steele, and Axel Schweiger (2008). "What Drove the Dramatic Retreat of Arctic Sea Ice during Summer 2007?" *Geophysical Research Letters* 35: L11505; Stroeve, J., M. Serreze, S. Drobot, S. Gearheard, M. Holland, J. Maslanik, W. Meier, and T. Scambos (2008). "Arctic Sea Ice Extent Plummets in 2007." *EOS* 89: No.2 13–20.

7. Lindsay, R. W., and J. Zhang (2005). "The Thinning of Arctic Sea Ice, 1988–2003: Have We Passed a Tipping Point?" *Journal of Climate* 18: 4879–4894; Holland, Marika M., Cecilia M. Bitz, and Bruno Tremblay (2006). "Future Abrupt Reductions in the Summer Arctic Sea Ice." *Geophysical Research Letters* 33: L23503.

8. Gillet, Nathan P., Dáithí A. Stone, Peter A. Stott, Toru Nozawa, Alexey Y. Karpechko, Gabriele Hegerl, Michael F. Wehner, and Philip D. Jones, (2008). "Attribution of Polar Warning to Human Influence." *Nature Geoscience* 1: 750–754.

9. Stroeve, Julienne, Marika M. Holland, Walt Meier, Ted Scambos, and Mark Serreze (2007). "Arctic Sea Ice Decline: Faster than Forecast." *Geophysical Research Letters* 34: L09501.

10. Serreze, Mark C. and Jennifer A. Francis (2006). "The Arctic Amplifica-
    tion Debate." *Climatic Change* 76: 241–264; Serreze, Mark C., and Julienne
    C. Stroeve (2008). "Standing on the Brink." *Nature Reports Climate Change.*

11. Wang, Muyin, and James E. Overland. "A Sea Ice Free Summer Arctic
    within 30 years?" *Geophysical Research Letters* 36: L07502.

12. Hansen, James, Makiko Sato, Pushker Kharecha, David Beerling, Robert
    Berner, Valerie Masson-Delmotte, Mark Pagani, Maureen Raymo, Dana
    L. Royer, and James C. Zachos (2008). "Atmospheric CO2: Where Should
    Humanity Aim?" *The Open Atmospheric Science Journal* 2: 217–231; Beerling,
    David (2007). *The Emerald Planet: How Plants Changed Earth's History.* Oxford:
    Oxford University Press.

13. Washington, W. M., R. Knutti, G. A. Meehl, H. Teng, C. Tebaldi, D. Law-
    rence, L. Buja, and W. G. Strand (2009). "How Much Climate Change Can
    be Avoided by Mitigation?" *Geophysical Research Letters* 36: L08703.

CHAPTER SEVEN: WHO OWNS THE ARCTIC?

1. Commission of the European Communities (2008). *The European Union
   and the Arctic Region.* Communication from the Commission to the European Parliament and the Council. Brussels: 20, November 2008. The EU's
   interest in the Arctic is driven by its dependence on Arctic energy from
   Norway and Russia. The logic is spelled out in this paper from Norway's
   Foreign Minister. Store, Jonas Gahr (2006). *The Emergence of the Barents Sea
   as a Petroleum Province Implications for Norway and Europe.* EPC Policy Briefing. Brussels, October 10, 2006.

2. The White House (2009). Arctic Region Policy. National Security Presidential Directive/NSPD-66, Homeland Security Presidential Directive/
   HSPD-25. January 9, 2009.

3. "Basics of the State Policy of the Russian Federation in the Arctic for the
   Period until 2020." *Rossiyskaya Gazeta* March 2009. The *Russian Gazette* is
   the Russian State newspaper, and acts of state come into force on publication within it.

4. *The Ilulissat Declaration*, Arctic Ocean Conference, Ilulissat, Greenland, May
   27–29, 2008.

5. European Parliament Resolution on Arctic Governance, Brussels, October
   9, 2008.

6. World Wildlife Fund (2008). *A New Sea: The Need for a Regional Agreement on
   Management and Conservation of the Arctic Marine Environment.*

7. Srivastava, S. P., H. Varma, and R. Macnab. "The Lomonosov, Alpha and
   Mendeleev Ridges: Tectonic Scenarios in the Arctic Ocean and the Test of

Appurtenance in UNCLOS Article 76." Kaminsky, V. D., V. A. Poselov, V. Y. Glebovsky, A. V. Zayonchek, and V. V. Butsenko. "Geophysical and Geological Study of the Transition Zone between the Mendeleev Rise and the Adjacent Siberian Shelf: Preliminary Results."

8. Berkman, P. A., and O. R. Young (2009). "Governance and Environmental Change in the Arctic Ocean." *Science* 324: 339–340. Provides an interesting perspective on how "the overlying water column and sea surface of the central Arctic can remain an undisputed international arena in which the interests of Arctic and non-Arctic states alike play a role in the development of effective governance." See also P. A. Berkman (2009). "North Pole As a Pole of Peace," *The Circle* no. 1, pub WWF International Arctic Programme, Oslo.

9. Tan, Alan Khee-Jin (2005). *Vessel-Source Marine Pollution: The Law and Politics of International Regulation.* Cambridge: Cambridge University Press.

10. Jensen, Øystein (2007). *The IMO Guidelines for Ships Operating in Arctic Ice-covered Waters. From Voluntary to Mandatory Tool for Navigation Safety and Environmental Protection?* Norway: The Fridtjof Nansen Institute.

11. Corell, Robert (2009). "Climate Change and the Arctic: New Frontiers of National Security. An Overview of the Science and National Security Interests of Climate Change in the Arctic." Testimony to the Committee on Foreign Relations, U.S. House of Representatives, March 25, 2009. See also www.arcticgovernance.org.

12. McGlade, Jacqueline (2007). "The Arctic Environment—Why Europe Should Care." A speech to the Arctic Frontiers Conference, Tromsø, January 23, 2007.

13. "Nunavut Premier Wants EU Barred from Arctic Council," Canadian Broadcasting Corporation (CBC) online news Wednesday, April 15, 2009; "EU Not Awarded Permanent Observer Status in Arctic Council," Inge S. Rasmussen, April 27, 2009, *Sermitsiaq,* Greenland's national newspaper.

14. Statement issued by Inuit Leaders at the Inuit Leaders' Summit on Arctic Sovereignty in Kuujjuaq, November 6–7, 2008.

CHAPTER 8: THE STRANGE CASE OF SVALBARD

1. Svalbard Satellite Station is owned by Kongsberg Satellite Services.

2. EISCAT (European Incoherent Scatter) studies the interaction between the sun and the earth in the magnetosphere and the atmosphere, interactions that give rise to the aurora.

3. Humlum, Ole. *A Geographical-Historical Outline of Svalbard.* Oslo: University of Oslo.

4. Scoresby, William (1820). *An Account of the Arctic Regions with a History and a Description of the Northern Whale-Fishery*. Edinburgh: Archibald Constable and Co.

5. Arlov, Thor B. (1989). *A Short History of Svalbard*. Oslo: Norsk Polarinstitutt, page 3.

6. Arctic Marine Shipping Assessment, page 39.

7. Captain Jonas Poole (1610). Cited in Longyearbyen Museum.

8. Henrat, Philippe (1984). "French Naval Operations in Spitsbergen During Louis XIV's Reign." *Arctic* 37 no. 4: 544–551.

9. Prestvold, Kristin (2001). *Smeerenburg Gravneset: Europe's First Oil Adventure*. Longyearbyen, Norway: Governor of Svalbard, Environmental Section.

10. Treaty among Norway, the United States of America, Denmark, France, Italy, Japan, the Netherlands, Great Britain, Ireland, the British overseas dominions, and Sweden concerning Spitsbergen, signed in Paris, February 9, 1920.

CHAPTER NINE: TROUBLE AT THE TOP

1. An excellent book on the animals of the Arctic is Blix, Arnoldus Schytte (2005). *Arctic Animals and Their Adaptations to Life on the Edge*. Trondheim: Tapir Academic Press. A useful guide for marine mammals is National Audubon Society (2002). *Guide to Marine Mammals of the World*. New York: Alfred A. Knopf.

2. Stirling, Ian (1998). *Polar Bears*. Ann Arbor: University of Michigan Press, page 118.

3. In note 2 above, page 119.

4. In note 2 above, page 119.

5. In note 2 above, page 118.

6. See papers by M. A. Ramsay, Department of Biology, University of Saskatchewan, Saskatoon, Canada.

7. Regehr, E. V., N. J. Lunn, S. C. Amstrup, and I. Stirling (2007). "Effects of Earlier Sea Ice Breakup on Survival and Population Size of Polar Bears in Western Hudson Bay." *Journal of Wildlife Management* 71, no. 8: 2673.

8. Durner, George, David Douglas, Ryan Nielson, Steven Amstrup, and Trent McDonald (2007). "Predicting the Future Distribution of Polar Bear Habitat in the Polar Basin from Resource Selection Functions Applied to Twenty-First-Century General Circulation Model Projections of Sea Ice." U.S Geological Survey.

9. Stirling, Ian, and Thomas G. Smith (2004). "Implications of Warm Temperatures and an Unusual Rain Event for the Survival of Ringed Seals

on the Coast of Southeastern Baffin Island." *Arctic* 57: 59–67; Stirling, Ian (2005). "Reproductive Rates of Ringed Seals and Survival of Pups in Northwestern Hudson Bay, Canada, 1991–2000." *Polar Biology* 28: 381–387; Ferguson, Steven H., Ian Stirling, and Philip McLoughlin (2005). "Climate Change and Ringed Seal (*Phoca hispida*) Recruitment in Western Hudson Bay." *Marine Mammal Science* 21: 121–135.

10. Kovacs, Kit M., and Christian Lydersen (2008). "Climate Change Impacts on Seals and Whales in the North Atlantic Arctic and Adjacent Shelf Seas." *Science Progress* 91: 117–150.

11. Population estimate 2.3 to 7 million. Huntington, Henry P., et al. "Arctic Flora and Fauna: Status and Conservation. A report from the Program for the Conservation of Arctic Flora and Fauna." *Status and Trends in Species and Populations*, page 220.

12. In note 11 above, page 220.

13. Cooper, Lee W., Carin J. Ashjian, Sharon L. Smith, Louis A. Codispoti, Jacqueline M. Grebmeier, Robert G. Campbell, and Evelyn B. Sherr (2006). "Rapid Seasonal Sea-Ice Retreat in the Arctic Could Be Affecting Pacific Walrus (*Odobenus rosmarus divergens*) Recruitment." *Aquatic Mammals* 32, no. 1: 98.

14. Seaver, Kirsten A. (1996). *The Frozen Echo: Greenland and the Exploration of North America ca. A.D. 1000–1500*. Stanford: Stanford University Press.

15. See note 11 above, page 200.

16. A good and entertaining book about the Cook Inlet beluga is Lord, Nancy (2007). *Beluga Days: Tracking the Endangered White Whale*. Seattle: The Mountaineers Books.

17. See note 11 above, page 219.

18. Laidre, Kristin L., Ian Stirling, Lloyd F. Lowry, Øystein Wiig, Mads Peter Heide-Jørgensen, and Steven H. Ferguson (2008). "Quantifying the Sensitivity of Arctic Marine Mammals to Climate-Induced Habitat Change." *Ecological Applications* 18: S97–S125.

CHAPTER TEN: THE BOTTOM OF THE WEB

1. A good overall look at the life at the bottom of the food chain, written principally for specialists, is Thomas, D.N., et al. (2008). *The Biology of Polar Regions*. Oxford: Oxford University Press.

2. Weissenberger, Jürgen, Rolf Gradinger, et al. (1992). "Sea Ice: A Cast Technique to Examine and Analyze Brine Pockets and Channel Structure." *Limnology and Oceanography* 37: 179–183.

3. Krembs. C., R. Gradinger, and M. Spindler (2000). "Implications of Brine

Channel Geometry and Surface Area for the Interaction of Sympagic Organisms in Arctic Sea Ice." *Journal of Experimental Marine Biology and Ecology* 243: 55–80.

4. Gradinger, Rolf R. (2001). "Adaptation of Arctic and Antarctic Ice Metazoa to Their Habitat." *Zoology* 104: 339–345.

5. Arrigo, Kevin R., Gert van Dijken, and Sudeshna Pabi (2008). "Impact of a Shrinking Arctic Ice Cover on Marine Primary Production." *Geophysical Research Letters* 35: L19603.

6. Bluhm, Bodil, and Rolf Gradinger (2008). "Regional Variability in Food Availability for Arctic Marine Mammals." *Ecological Applications* 18: S77–S96.

7. Grebmeier, Jacqueline M., James E. Overland, Sue E. Moore, Ed V. Farley, Eddy C. Carmack, Lee W. Cooper, Karen E. Frey, John H. Helle, Fiona A. McLaughlin, and S. Lyn McNutt (2006). "A Major Ecosystem Shift in the Northern Bering Sea." *Science* 311: 1461.

8. Frey, Darcy. "George Divoky's Planet." *New York Times,* January 6, 2002.

9. You can see much of Divoky's research at the Web site of the Friends of Cooper Island, "Monitoring Climate Change with Arctic Seabirds," cooperisland. org.

10. The video was produced for Polar-Palooza. It is called "George Divoky: The Bird-Watcher Who Saw the Future," and can be found on YouTube.

CHAPTER ELEVEN: INVADERS FROM THE SOUTH

1. See Chapter Nine, note 10. Higdon, Jeff and Steven Ferguson (2009). "Loss of Arctic Sea Ice Causing Punctuated Change in Sightings of Killer Whales over the Past Century." *Ecological Applications* 1: 1365.

2. Berge, Jørgen, et al. (2005) "Ocean Temperature Oscillations Enable Reappearance of Blue Mussels (*Mytilus edulis*) in Svalbard after a 1000 Year Absence." *Marine Ecology Progress Series* 303: 167.

3. Vermeij, Geerat J., and Peter D. Roopnarine (2008). "The Coming Arctic Invasion." *Science* 321: 780–781.

4. Greene, Charles H, Andrew J. Pershing, Thomas M. Cronin, and Nicole Ceci (2008). "Arctic Climate Change and Its Impacts on the Ecology of the North Atlantic." *Ecology* 89: S24–S38.

5. Stokke, Olav Schram (2003). "Management of Shared Fish Stocks in the Barents Sea." *Papers Presented at the Norway-FAO Expert Consultation on the Management of Shared Fish Stocks Bergen, Norway, 7–10 October 2002,* FAO Fisheries Report No. 695 Supplement, Rome: 180–191.

6. Benoit, D., Y. Simard, and L. Fortier (2008). "Hydroacoustic Detection of

Large Winter Aggregations of Arctic Cod (*Boreogadus saida*) at Depth in Ice-Covered Franklin Bay (Beaufort Sea)." *Journal of Geophysical Research* 113: C06S90.

7. Moore, Sue E., and Henry P. Huntington (2008). "Arctic Marine Mammals and Climate Change Impacts and Resilience." *Ecological Applications* 18: S157–S165.

CHAPTER TWELVE: THE BATTLE FOR ARCTIC OIL

1. See also Arctic Monitoring and Assessment Programme (AMAP). *Arctic Oil and Gas,* 2007.
2. Hess, Bill (1999). *Gift of the Whale: The Inupiat Bowhead Hunt, a Sacred Tradition.* Seattle: Sasquatch Books.
3. The Arctic Energy Summit Technology Conference, Anchorage, Alaska, October 15–18, 2007.
4. "Economic Analysis of Future Offshore Oil and Gas Development: Beaufort Sea, Chukchi Sea, and North Aleutian Basin." Prepared for Shell Exploration and Production by Northern Economics, March 2009.
5. Niini, Mikko, Sergey Kaganov, and Robert D. Tustin (2007). *Development of Arctic Double Acting Shuttle Tankers for the Prirazlomnoye Project.* TSCF 2007 Shipbuilders Meeting.
6. The Oslo-based environmental group Bellona has produced a report on the dangers of offshore oil and gas development in the Russian Arctic by Nina Lesikhina, Irina Rudaya, and Anna Kireeva. See www.bellona.org.
7. Bid prices and conditions are listed on the Web site of Indian and Northern Affairs Canada, www.ainc-inac.gc.ca.

CHAPTER THIRTEEN: HOW FAR CAN OIL GO?

1. The logbook of the Arctic Coring Expedition 2004 can be found at the Web site of the University of Bremen Center for Marine Environmental Sciences, www.marum.de/en/Expedition_Logbook.html.
2. Stoll, Heather M. (2006). "Climate Change: The Arctic Tells Its Story." *Nature* 441: 579–581.
3. Paulin, Michael J. *Arctic Offshore Technology Assessment of Exploration and Production Options for Cold Regions of the U.S. Outer Continental Shelf.* IMV Projects Atlantic Inc., St. John's, Newfoundland, Canada for the U.S. Minerals Management Service. February 6, 2008; see also "Oil and Gas Technologies for the Arctic and Deepwater." U.S. Congress Office of Technology Assessment (1985).

CHAPTER FOURTEEN: TOO MANY SHIPS, TOO SOON?

1. Arctic Marine Shipping Assessment (AMSA) 2009 Report, Arctic Council, April 2009.
2. *Opening the Arctic Seas: Envisioning Disaster & Framing Solutions.* March 18–20, 2008. The University of New Hampshire.
3. A gripping account is found in Jeffers, H. Paul (2006). *Burning Cold: The Cruiseship Prinsendam and the Greatest Sea Rescue of All Time.* Motorhead Books.
4. Bambulyak, Alexei, and Bjørn Frantzen (2008). *Oil Transport from the Russian Part of the Barents Region.* Report of the Norwegian Barents Secretariat. Akvaplan-niva AS.
5. *Polar Icebreakers in a Changing World: An Assessment of U.S. Needs.* National Academy of Engineering 2007.
6. Niini, Mikko, Sergey Kaganov, and Robert D. Tustin (2007). *Development of Arctic Double Acting Shuttle Tankers for the Prirazlomnoye Project.* TSCF 2007 Shipbuilders Meeting.
7. See note 1 above, page 82.
8. There are numerous books on the Northwest Passage and the explorers who died trying to find their way through it. The first success was in the ship *Gjøa*, which now sits outside the museum in Oslo built to house the *Fram*. Anundsen's own account of the voyage is strongly recommended. Amundsen, Roald (2006). *The North West Passage: Being the Record of a Voyage of Exploration of the Ship* Gjøa. Elibron Classics.
A really excellent account of the various expeditions including those to the pole is Berton, Pierre (2000). *Arctic Grail: The Quest for the Northwest Passage and the North Pole, 1818–1909.* New York: The Lyon Press.
There are many books on the Franklin expedition and the continuing mystery of where the two ships sank and exactly what happened to the crew as they plodded south across the ice in the hope of reaching settlements far away in Canada. Whether they resorted to cannibalism on route remains controversial. Much depends on the accounts collected from Inuit living in the area. There are two books that take a unique perspective. One looks at John Rae, the explorer who first found traces of the missing expedition and brought back news that Victorian Britain did not want to hear. McGoogan, Ken (2001). *Fatal Passage: The Story of John Rae, The Arctic Hero Time Forgot.* New York: Carroll & Graf. The other looks in great detail at the Inuit testimony and is a wonderful book about this fascinating mystery. Woodman, David C. (1992) *Unravelling the Franklin Mystery: Inuit Testimony.* Montreal, Canada: McGill-Queen's University Press.

9. See note 1 above, page 21.

10. "First Commercial Ship Sails through Northwest Passage," Canadian Broadcasting Corporation online news service, November 28, 2008.

11. *Decision of the Commissioner of Maritime Affairs of the Republic of Liberia and the Report of Investigation in the Matter of Sinking of Passenger Vessel MV* Explorer *23 November 2007 in the Bransfield Strait near the South Shetland Islands.* Bureau of Maritime Affairs, March 26, 2009, Monrovia, Liberia.

12. See note 1, pages 160 and 166.

13. Norwegian Atlantic Committee (2006). "Developments in Arctic Shipping." Focus North, No. 8; The Company of Master Mariners of Canada, the Marine Affairs Program of Dalhousie University; Lloyd's Register, North America (2007). *Canadian Arctic Issues in a Changing Climate.*

14. See note 1, page 33.

15. See note 1, page 44.

16. See note 1, page 44.

17. These include the International Northern Sea Route Programme (IN-SROP), which ran from 1993 to 1999 and the Arctic Operational Platform (ARCOP).

18. "Breaking The Ice. Arctic Development and Maritime Transportation: Prospects of the Transarctic Route—Impact and Opportunities," for Arctic Council's Arctic Marine Shipping Assessment, March 27–28, 2007, Akureyri, Iceland.

19. See note 1, page 105.

20. Niini, M., M. Arpiainen, and R. Kiili (2006). "Arctic Shuttle Container Link from Alaska, U.S. to Europe." Arker Arctic Technology Inc. Report K–63.

21. Committee on Cumulative Environmental Effects of Oil and Gas Activities on Alaska's North Slope, National Research Council of the National Academies (2003). "Cumulative Environmental Effects of Oil and Gas Activities on Alaska's North Slope." Washington, D.C.: The National Academies Press.

22. The full story is found in *Marine Accident Brief Accident No.: DCA-05-MM-008; Bulk Carrier M/V* Selendang Ayu (2006). National Transportation Safety Board, Washington, D.C.

23. World Wildlife Fund (2007). "Oil Spill: Response Challenges in Arctic Waters."

24. Shell Exploration and Production Co. (2007). *Shell's Beaufort Sea Exploratory Drilling Program Oil Spill Response.*

25. Camphuysen, Kees. *Chronic Oil Pollution in Europe: A Status Report.* The International Fund for Animal Welfare, IFAW.org.

26. See the Circumpolar Vulnerability Maps prepared by Hein Rune Skjoldal of the Institute of Marine Research, Bergen, Norway, for Chapter 6 of the Arctic Council/AMAP Assessment of Oil and Gas Activities in the Arctic.

27. Testimony by Mead Treadwell to the House Committee on Foreign Affairs (2009). *Climate Change and the Arctic: New Frontiers of National Security*, March 25, 2009.

CHAPTER FIFTEEN: THE ARCTIC'S REVENGE

1. Stroeve, J.C., et al. (2008). "Emerging Arctic Amplification as Seen in the NCEP/NCAR Reanalysis." *EOS, Transactions American Geophysical Union* 89: Abstract C41B–0502.

2. Francis, Jennifer A., Weihan Chan, Daniel J. Leathers, James R. Miller, and Dana E. Veron (2009). "Winter Northern Hemisphere Weather Patterns Remember Summer Arctic Sea-Ice Extent." *Geophysical Research Letters* 36: L07503.

3. Lawrence, David M., Andrew G. Slater, Robert A. Tomas, Marika M. Holland, and Clare Deser (2008). "Accelerated Arctic Land Warming and Permafrost Degradation During Rapid Sea Loss." *Geophysical Research Letters* 35: L11506.

4. Sturm, Matthew, Josh Schimel, Gary Michaelson, Jeffrey M. Welker, Steven F. Oberbauer, Glen E. Liston, Jace Fahnestock, and Vladimir E. Romanovsky (2005). "Winter Biological Processes Could Help Convert Arctic Tundra to Shrubland." *BioScience* 55: 17–26; (2008). Chapin III, F. S., M. Sturm, M. C. Serreze, J. P. McFadden, J. R. Key, A. H. Lloyd, A. D. McGuire, T. S. Rupp, A. H. Lynch, J. P. Schimel, J. Beringer, W. L. Chapman, H. E. Epstein, E. S. Euskirchen, L. D. Hinzman, G. Jia, C.-L. Ping, K. D. Tape, C. D. C. Thompson, D. A. Walker, and J. M. Welker (2005). "Role of Land-Surface Changes in Arctic Summer Warming." *Science* 310: 657–660.

5. Plant growth may initially help offset carbon lost as permafrost thaws, but it will be unable to keep up, and thawing Arctic soil could add a billion tons of carbon per year to the atmosphere. See Vogel, Jason G., Kathryn G. Crummer, Hanna Lee, James O. Sickman, T. E. Osterkamp, Edward A. G. Schuur (2009). "The Effect of Permafrost Thaw on Old Carbon Release and Net Carbon Exchange from Tundra." *Nature* 459: 556–559.

6. Tedesco, M., X. Fettweis, M. van den Broeke, R. van de Wal, and P. Smeets (2008). "Melting and Surface Mass Balance over the Greenland Ice Sheet from Satellite Data, Model Results and Ground Measurements During IPY: Extreme Events and Updated Trends." *EOS, Transactions American*

Geophysical Union 89, no. 53, Fall Meet. Suppl., Abstract U23F–01; Tedesco, M., X. Fettweis, M. van den Broeke, R. van de Wal, and P. Smeets (2008). "Extreme Snowmelt in Northern Greenland During Summer 2008." *EOS, Transactions American Geophysical Union* 89, no. 41.

7. Tedesco, M. (2007). "A New Record in 2007 for Melting in Greenland." *EOS, Transactions American Geophysical Union* 88, no. 39.

8. A well-illustrated and informative book on the Ilulissat World Heritage Site and the history of the glacier is Bennike, Ole, et al. (2004). *Ilulissat Icefjord.* Copenhagen: Geological Survey of Denmark and Greenland.

9. Rignot, Eric, and Pannir Kanagaratnam (2006). "Changes in the Velocity Structure of the Greenland Ice Sheet." *Science* 311: 986–990.

10. Rignot, E., J. E. Box, E. Burgess, and E. Hanna (2008). "Mass Balance of the Greenland Ice Sheet from 1958–2007." *Geophysical Research Letters* 35: L20502.

11. Joughin, Ian (2008). "Seasonal Speedup Along the Western Flank of the Greenland Ice Sheet." *Science* 320: 781.

12. Holland, David M., Robert H. Thomas, Brad De Young, Mads H. Ribergaard, and Bjarne Lyberth (2008). "Acceleration of Jakobshavn Isbræ Triggered by Warm Subsurface Ocean Waters." *Nature Geoscience* 1: 659.

13. Shepherd, Andrew, and Duncan Wingham (2007). "Recent Sea-Level Contributions of the Antarctic and Greenland Ice Sheets." *Science* 315: 1529–1532; Dowdeswell, Julian A. (2006). "The Greenland Ice Sheet and Global Sea-Level Rise." *Science* 311: 963–964.

14. Pfeffer, W. T., J. T. Harper, and S. O'Neel (2008). "Kinematic Constraints on Glacier Contributions to 21st-Century Sea-Level Rise." *Science* 321: 1340–1343.

15. Walter, K. M., S. A. Zimov, J. P. Chanton, D. Verbyla, and F. S. Chapin III (2006). "Methane Bubbling from Siberian Thaw Lakes as a Positive Feedback to Climate Warming." *Nature* 443: 71–75; Walter, K. M., M. E. Edwards, G. Grosse, S. A. Zimov, and F. S. Chapin III (2007). "Thermokarst Lakes as a Source of Atmospheric CH_4 During the Last Deglaciation." *Science* 318: 633–636.

16 Petrenko, Vasilii (2009). *Science* 324: 506. "CH_4 Measurements in Greenland Ice: Investigating Last Glacial Termination CH_4 Sources."

17. Archer, D. (2007). "Methane hydrate stability and anthropogenic climate change." *Biogeosciences* 4: 521–544; Archer, David (2009). *The Long Thaw: How Humans are Changing the Next 100,000 Years of Earth's Climate.* Princeton: Princeton University Press. An excellent book describing the real threat—not so much the immediate change to climate but the long-term impacts we are having on the planet.

18. Romanovsky, V. E., et al. (2008). "Change in permafrost in Northern Eurasia." *EOS, Transactions American Geophysical Union* 89: Abstract GC52A-04.

19. Lawrence, David M., and Andrew G. Slater (2005). "A Projection of Severe Near-Surface Permafrost Degradation During the Twenty-first Century." *Geophysical Research Letters* 32: L24401.

20. Lawrence, David M., Andrew G. Slater, Vladimir E. Romanovsky, and Dmitry J. Nicolsky (2008). "Sensitivity of a model projection of near-surface permafrost degradation to soil column depth and representation of soil organic matter." *Journal of Geophysical Research* 113: F02011.

21. Zimov, Sergey A., Edward A. G. Schuur, and F. Stuart Chapin III (2006). "Permafrost and the Global Carbon Budget." *Science* 312: 1612–1613.

22. Khvorostyanov, D. V., P. Ciais, G. Krinner, and S. A. Zimov (2008). "Vulnerability of East Siberia's Frozen Carbon Stores to Future Warming." *Geophysical Research Letters* 35: L10703.

23. Greene, Charles H, Andrew J. Pershing, Thomas M. Cronin, and Nicole Ceci (2008). "Arctic Climate Change and Its Impacts on the Ecology of the North Atlantic." *Ecology* 89: S24–S38.

24. Zimov, Sergey, et al. (2008). "Carbon Storage in Frozen Loess and Soils of the Mammoth Tundra-Steppe Biome." *EOS, Transactions American Geophysical Union* 89: Abstract B23B-0429. Zimov, Sergey (2005). "Pleistocene Park." *Science* 308: 796.

CHAPTER SIXTEEN: BLACK AND WHITE

1. Nordenskiöld, Adolf Erik (1879). *The Arctic Voyages of Nordenskiöld, 1858–1879.* Macmillan and Co.; Nordenskiöld, Adolf Erik (2008). *Voyage of the Vega.* Hesperides Press.

2. Nordenskiöld, Adolf Erik (1883). "Nordenskiöld on the Inland Ice of Greenland." *Science* 2, no. 44: 732.

3. Garrett, Timothy J., and Lisa L. Verzella (2008). "Looking Back: An Evolving History of Arctic Aerosols." *Bulletin of the American Meteorological Society* 89, no. 3: 299.

4. McConnell, Joseph R., Ross Edwards, Gregory L. Kok, Mark G. Flanner, Charles S. Zender, Eric S. Saltzman, J. Ryan Banta, Daniel R. Pasteris, Megan M, Carter, and Jonathan D. W. Kahl (2007). "Twentieth-Century Industrial Black Carbon Emissions Altered Arctic Climate Forcing." *Science* 317: 1381; Alley, Richard B. (2000). *The Two-Mile Time Machine: Ice Cores, Abrupt Climate Change, and Our Future.* Princeton: Princeton University Press; The classic account of coring ice and reconstructing past climate.

Turney, Chris (2008). *Ice, Mud and Blood: Lessons from Climates Past.* New York: Macmillan. Another interesting book.

5. Garrett, Timothy J., and Chuanfeng Zhao (2006). "Increased Arctic Cloud Long-Wave Emissivity Associated with Pollution from Mid-Latitudes." *Nature* 440: 787–789.

6. "A Discussion of Near-term Strategies for Slowing Warming in the Arctic: Short-term Pollutants and Arctic Warming." Oslo, Norway, November 7–8, 2007.

7. Zender, Charles S. (2007). "Arctic Climate Effects of Black Carbon." U.S. House of Representatives. Written testimony to the Oversight and Government Reform Committee; Hansen, James, and Larissa Nazarenko (2004). "Soot Climate Forcing via Snow and Ice Albedos." *PNAS* 101, no. 2: 423–428.

8. Flanner, M. G., Charles S. Zender, P. G. Hess, N. M. Mahowald, T. H. Painter, V. Ramanathan, and P. J. Rasch (2008). "Springtime Warming and Reduced Snow Cover from Carbonaceous Particles." *Atmospheric Chemistry and Physics Discussion* 8: 19819–19859.

9. Quinn, P. K., T. S. Bates, E. Baum, N. Doubleday, A. M. Fiore, M. Flanner, A. Fridlind, T. J. Garrett, D. Koch, S. Menon, D. Shindell, A. Stohl, and S. G. Warren (2007). "Short-Lived Pollutants in the Arctic: Their Climate Impact and Possible Mitigation Strategies." *Atmospheric Chemistry and Physics Discussion* 7: 15669–15692; Quinn, P. K., et al. (2008). "The Impact of Short-Lived Pollutants on Arctic Climate." *Arctic Monitoring and Assessment Programme Technical Report No 1.*

10. Robert Watson, a former head of the Intergovernmental Panel on Climate Change (IPCC) chaired a session on aerosols and Arctic warming in Poznan.

11. Quinn, P. K., et al. (2008). "The Impact of Short-Lived Pollutants on Arctic Climate." *Arctic Monitoring and Assessment Programme Technical Report No 1*; Bluestein, J., J. Rackley, and E. Baum (2008). "Sources and Mitigation Opportunities to Reduce Emissions of Short-term Arctic Climate Forcers." *Arctic Monitoring and Assessment Programme Technical Report No 2*; Reducing Black Carbon May Be Faster Strategy for Slowing Climate Change (2008) IGSD/INECE Climate Briefing Note.

12. Stohl, A., T. Berg, J. F. Burkhart, A. M. Fjæraa, C. Forster, A. Herber, Ø. Hov, C. Lunder, W. W. McMillan, S. Oltmans, M. Shiobara, D. Simpson, S. Solberg, K. Stebel, J. Ström, K. Tørseth, R. Treffeisen, K. Virkkunen, and K. E. Yttri (2007). "Arctic Smoke—Record High Air Pollution Levels in the European Arctic Due to Agricultural Fires in Eastern Europe in Spring

2006." *Norway Atmospheric Chemistry and Physics* 7: 511–534.; Law, Kathy S. and Andreas Stohl (2007). "Arctic Air Pollution: Origins and Impacts." *Science* 315: 1537.

13 Warneke, C., R. Bahreini, J. Brioude, C. A. Brock, J. A. de Gouw, D. W. Fahey, K. D. Froyd, J. S. Holloway, A. Middlebrook, L. Miller, S. Montzka, D. M. Murphy, J. Peischl, T. B. Ryerson, J. P. Schwarz, J. R. Spackman, and P. Veres. (2009) "Biomass Burning in Siberia and Kazakhstan as an Important Source for Haze over the Alaskan Arctic in April 2008." *Geophysical Research Letters* 36: L02813.

14. Pettus, Ashley. (2009) *Agricultural Fires and Arctic Climate Change*. Clean Air Task Force.

15. Nordenskiöld, Adolf Erik (1883). "Nordenskiöld on the Inland Ice of Greenland." *Science* 2, no. 44: 733.

CHAPTER SEVENTEEN: THE FUTURE OF THE ARCTIC

1. "Glacial El Dorado: Spitsbergen" (2001). Moscow: Punta.

2. The company is called "Xconomy.com."

3. Arnason, Ragnar (2008). "Climate Change and Fisheries: Assessing the Economic Impact in Iceland and Greenland." *Natural Resource Modeling* 20: 163–197

4. Brigham, Lawson (2007). "Thinking about the Arctic's Future: Scenarios for 2040." *The Futurist*. September/October; Brunstad, Bjørn. "Arctic Shipping 2030"; *From Russia with Oil, Stormy Passage or Arctic Great Game?* commissioned by Nor-Shipping from Econ, Oslo.

INDEX